Lena Ivanova

Nitrogen Containing III-V Semiconductor Surfaces and Nanostructures

AF004681

Lena Ivanova

Nitrogen Containing III-V Semiconductor Surfaces and Nanostructures

Studied by Scanning Tunneling Microscopy and Spectroscopy

Südwestdeutscher Verlag für Hochschulschriften

Impressum / Imprint
Bibliografische Information der Deutschen Nationalbibliothek: Die Deutsche Nationalbibliothek verzeichnet diese Publikation in der Deutschen Nationalbibliografie; detaillierte bibliografische Daten sind im Internet über http://dnb.d-nb.de abrufbar.
Alle in diesem Buch genannten Marken und Produktnamen unterliegen warenzeichen-, marken- oder patentrechtlichem Schutz bzw. sind Warenzeichen oder eingetragene Warenzeichen der jeweiligen Inhaber. Die Wiedergabe von Marken, Produktnamen, Gebrauchsnamen, Handelsnamen, Warenbezeichnungen u.s.w. in diesem Werk berechtigt auch ohne besondere Kennzeichnung nicht zu der Annahme, dass solche Namen im Sinne der Warenzeichen- und Markenschutzgesetzgebung als frei zu betrachten wären und daher von jedermann benutzt werden dürften.

Bibliographic information published by the Deutsche Nationalbibliothek: The Deutsche Nationalbibliothek lists this publication in the Deutsche Nationalbibliografie; detailed bibliographic data are available in the Internet at http://dnb.d-nb.de.
Any brand names and product names mentioned in this book are subject to trademark, brand or patent protection and are trademarks or registered trademarks of their respective holders. The use of brand names, product names, common names, trade names, product descriptions etc. even without a particular marking in this work is in no way to be construed to mean that such names may be regarded as unrestricted in respect of trademark and brand protection legislation and could thus be used by anyone.

Verlag / Publisher:
Südwestdeutscher Verlag für Hochschulschriften
ist ein Imprint der / is a trademark of
OmniScriptum GmbH & Co. KG
Heinrich-Böcking-Str. 6-8, 66121 Saarbrücken, Deutschland / Germany
Email: info@svh-verlag.de

Herstellung: siehe letzte Seite /
Printed at: see last page
ISBN: 978-3-8381-1183-4

Zugl. / Approved by: Berlin, TU, Diss., 2009

Copyright © 2009 OmniScriptum GmbH & Co. KG
Alle Rechte vorbehalten. / All rights reserved. Saarbrücken 2009

Abstract

III-V compound semiconductors are common materials for many semiconductor technologies and applications. In this work, in particular different nitrogen containing III-V semiconductor surfaces and nanostructures are studied, using scanning tunneling microscopy (STM) and spectroscopy.

In so-called diluted GaAsN layers single nitrogen atoms can be identified at the cross-sectional STM images as dark depressions within the arsenic rows of the (110) cleavage surface. On the base of height profiles from the images nitrogen-related features can be distinguished from other crystal defects such as impurities, vacancies, or interstitials. The measured density of states shows that nitrogen impurities lead to a splitting of the GaAs conduction band. The measured data are compared with theoretical calculation, showing an excellent agreement. Furthermore, it is shown that InAs/GaAsN quantum dots (QD)s capped with GaAs demonstrate much smaller sizes, as compared with InAs QDs grown on and capped with pure GaAs, while InAs/GaAs QDs capped with GaAsN do not exhibit any size changes. The incorporation of nitrogen with a nominal concentration of 9% into InAs/GaAs QDs leads to a rather strong dissolution and the formation of extended almost spherical nitrogen-free InGaAs QDs with a low indium content. It is shown that the InAsN/GaAs QD system demonstrates a trend to separate into InGaAs QDs and a GaAsN matrix.

Furthermore, for the $GaN(1\bar{1}00)$ cleavage surface of epitaxially grown GaN substrates it is found that both the nitrogen and gallium derived intrinsic dangling bond surface states are outside of the fundamental bulk band gap. Their band edges are both located at the $\bar{\Gamma}$ point of the surface Brillouin zone. The observed Fermi level pinning at 1.0 eV below the conduction band edge could be attributed to the high step and defect density at the surface, but not to intrinsic surface states. Moreover, in GaN wafers dislocations are found to form localized bunches of entangled non-parallel dislocation lines. Within these bunches uncharged perfect dislocations with $a/3\langle 11\bar{2}0\rangle$ Burgers vectors and negatively charged Shockley partial dislocations with $a/3\langle 1\bar{1}00\rangle$ Burgers vectors interconnected by a negatively charged stacking fault are found. Finally, an epitaxially grown silicon doping modulation structure is imaged. This modulation gives rise to a height modulation in constant-current STM images. The origin of the height modulation is traced to two contrast mechanisms, an electronic modulation of the band edge energies yielding a voltage dependent corrugation supported by a mechanical relaxation at the surface of the doping-induced strain yielding a voltage independent contrast modulation.

Zusammenfassung

III-V-Verbindungshalbleiter sind die häufig benutzten Materialien für viele Halbleitertechnologien und -anwendungen. In dieser Arbeit werden im speziellen verschiedene stickstoffhaltige III-V-Verbindungshalbleiter mittels Rastertunnelmikroskopie (RTM) und -spektroskopie untersucht.

In so genannten verdünnten GaAsN-Schichten können einzelne Stickstoffatome identifiziert werden. Sie erscheinen als dunkle Punkte in Arsenreihen der (110)-Spaltfläche. Auf der Grundlage von Höhenprofilen in den Bildern können Stickstoffatome von anderen Kristalldefekten wie Störstellen, Leerstellen und Zwischengitteratomen unterschieden werden. Die gemessene Zustandsdichte zeigt, dass die Stickstoffatome eine Aufspaltung des Leitungsbandes von GaAs hervorrufen. Der Vergleich der gemessenen Daten mit theoretischen Rechnungen zeigt eine hervorragende Übereinstimmung. Außerdem wird gezeigt, dass InAs/GaAsN Quantenpunkte (QP), die mit GaAs bedeckt sind, viel kleinere Größen aufweisen als InAs QP, die auf GaAs gewachsen und mit GaAs bedeckt sind, während InAs/GaAs QP, die mit GaAsN bedeckt sind, keine Größenänderung zeigen. Das Beimischen von Stickstoff mit einer nominellen Konzentration von 9% in die InAs/GaAs QP führt zu einer Auflösung und Ausbildung von ausgedehnten fast späherischen stickstofffreien InGaAs QP mit einem geringen Indiumgehalt. Es wird gezeigt, dass das InAsN/GaAs QP-System einen Trend aufweist, InGaAs QP und eine GaAsN-Matrix voneinander zu separieren.

Des Weiteren wird für die GaN$(1\bar{1}00)$-Spaltfläche von epitaktisch gewachsenen GaN-Substraten nachgewiesen, dass intrinsische Oberflächenzustände der freien Bindungen sowohl von Stickstoff- als auch von Galliumatomen außerhalb der fundamentalen Volumenbandlücke liegen. Ihre Bandkanten befinden sich am $\bar{\Gamma}$-Punkt der Oberflächen-Brillouin-Zone. Das Verankern der Fermienergie +1.0 eV unterhalb der Leitungsbandkante konnte auf die hohe Stufen- und Defektdichte auf der Oberfläche zurückgeführt werden, jedoch nicht auf die intrinsischen Oberflächenzustände. Außerdem wurde festgestellt, dass Versetzungen in GaN-Wafern mit nicht parallelen Versetzungslinien und gebündelt auftreten. In den Bündeln wurden ungeladene vollständige Versetzungen mit $a/3\langle 11\bar{2}0\rangle$ Burgersvektoren und negativ geladene Shockley-Partial-Versetzungen mit $a/3\langle 1\bar{1}00\rangle$ Burgersvektoren, die einen Stapelfehler beinhalten, festgestellt. Schließlich wurde eine Modulation der Siliziumdotierung abgebildet. Diese Modulation führt zu einer Höhenmodulation in RTM-Bildern, die im Konstantstrom-Modus aufgenommen wurden. Die Entstehung der Höhenmodulation wird auf zwei Kontrastmechanismen zurückgeführt, eine elektronische Modulation der Bandkantenenergien, die eine tunnelspannungsabhängige Welligkeit hervorruft, und eine mechanische dotierungsinduzierte Verspannungsrelaxation der Oberfläche, die eine tunnelspannungsunabhängige Kontrastmodulation hervorruft.

List of publications

Parts of this work:

Effect of nitrogen on the InAs/GaAs quantum dot shape
L. Ivanova, H. Eisele, A. Lenz, R. Timm, M. Dähne, O. Schumann, L. Geelhaar, and H. Riechert, phys. stat. sol. (c), submitted (2009).

Scanning tunneling microscopy on unpinned $GaN(1\bar{1}00)$ surfaces: Invisibility of valence band states
Ph. Ebert, L. Ivanova, and H. Eisele, Phys. Rev. B **80**, 085316 (2009).

Doping modulation in GaN imaged by cross-sectional scanning tunneling microscopy
H. Eisele, L. Ivanova, S. Borisova, M. Dähne, M. Winkelnkemper, and Ph. Ebert, Appl. Phys. Lett. **94**, 162110 (2009).

Structure and electronic properties of dislocations in GaN
Ph. Ebert, L. Ivanova, S. Borisova, H. Eisele, A. Laubsch, and M. Dähne, Appl. Phys. Lett. **94**, 062104 (2009).

Surface states and origin of the Fermi level pinning on non-polar $GaN(1\bar{1}00)$ surfaces,
L. Ivanova, S. Borisova, H. Eisele, M. Dähne, A. Laubsch, and Ph. Ebert, Appl. Phys. Lett. **93**, 192110 (2008).

Nitrogen-induced intermixing of InAsN quantum dots with the GaAs matrix
L. Ivanova, H. Eisele, A. Lenz, R. Timm, M. Dähne, O. Schumann, L. Geelhaar, and H. Richert, Appl. Phys. Lett. **92**, 203101 (2008).

Effects of strain and confinement on the emission wavelength of InAs quantum dots due to a $GaAs_{1-x}N_x$ capping layer
O. Schumann, S. Birner, M. Baudach, L. Geelhaar, H. Eisele, L. Ivanova, R. Timm, A. Lenz, S.K. Becker, M. Povolotskyi, M. Dähne, G. Abstreiter, and H. Riechert, Phys. Rev. B **71**, 245316 (2005).

Further publications:

Contrast mechanisms in cross-sectional scanning tunneling microscopy of GaSb/GaAs type-II nanostructures
R. Timm, R.M. Feenstra, H. Eisele, A. Lenz, L. Ivanova, M. Dähne, and E. Lenz,
J. Appl. Phys. **105**, 180504 (2009).

Limits of In(Ga)As/GaAs quantum dot growth
A. Lenz, H. Eisele, R. Timm, L. Ivanova, R.L. Sellin, H.-Y. Liu, M. Hopkinson,
U.W. Pohl, D. Bimberg, and M. Dähne, phys. stat. sol. (b), **246**, 717 (2009).

Self-organized formation of GaSb/GaAs quantum rings
R. Timm, H. Eisele, A. Lenz, L. Ivanova, G. Balakrishnan, D.L. Huffaker, and
M. Dähne, Phys. Rev. Lett., **101**, 256101 (2008).

Quantum ring formation and antimony segregation in GaSb/GaAs nanostructures
R. Timm, A. Lenz, H. Eisele, L. Ivanova, M. Dähne, G. Balakrishnan, D.L. Huffaker,
I. Farrer, and D.A. Ritchie, J. Vac. Sci. Technol. B **26**, 1492 (2008).

Structure of InAs quantum dots-in-a-well nanostructures
A. Lenz, H. Eisele, R. Timm, L. Ivanova, H.-Y. Liu, M. Hopkinson, U.W. Pohl, and
M. Dähne, Physica E **40**, 1988 (2008).

Structural investigation of hierarchically selfassembled GaAs/AlGaAs quantum dots
A. Lenz, R. Timm, H. Eisele, L. Ivanova, D. Martin, V. Voßebürger, A. Rastelli,
O.G. Schmidt, and M. Dähne, phys. stat. sol. (b) **243**, 3976 (2006).

Onset of GaSb/GaAs quantum dot formation
R. Timm, A. Lenz, H. Eisele, L. Ivanova, K. Pötschke, U.W. Pohl, D. Bimberg,
G. Balakrishnan, D. Huffaker, and M. Dähne, phys. stat. sol. (c) **3**, 3971 (2006).

Contents

1	**Introduction**	**11**
2	**Nitrogen containing III-V compound semiconductors**	**15**
	2.1 Crystal structure of group-III arsenides versus group-III nitrides	15
	2.2 Band bowing and band-anticrossing model in III-V nitrides	16
	2.2.1 Vegard law	16
	2.2.2 Diluted GaAsN	17
	2.2.3 Diluted GaInAsN	21
	2.3 Low-dimensional semiconductor structures	21
	2.3.1 Bound particles	22
	2.3.2 Formation of low-dimensional structures – Molecular beam epitaxy	23
	2.3.3 Strained layers	24
	2.4 Surface segregation	25
	2.4.1 Theory of segregation	25
	2.4.2 Intermixing at InAs/GaAs and GaAs/InAs interfaces	26
	2.5 Dislocations and stacking faults in the wurtzite structure	27
	2.5.1 Defects in crystals	27
	2.5.2 Dislocations and stacking faults in the wurtzite structure	28
	2.5.3 Movement of dislocations	31
	2.6 Growth of free-standing GaN substrates – Hydride vapor phase epitaxy	31
3	**Scanning tunneling microscopy and spectroscopy**	**35**
	3.1 Theory of tunneling	35
	3.1.1 One-dimensional tunneling	35
	3.1.2 Bardeen approach	37
	3.1.3 Tersoff-Hamann approximation	38

	3.2	STM imaging and spectroscopy	39
		3.2.1 STM operation modes	39
		3.2.2 Influence of parallel momentum	42
		3.2.3 Tip-induced band bending	43
	3.3	Plan view and cross-sectional STM	44
	3.4	STM on III-V semiconductors	45
		3.4.1 Contrast mechanisms	45
		3.4.2 Structural and electronic properties of the GaAs(110) surface . .	45
		3.4.3 Structural and electronic properties of non-polar GaN$(1\bar{1}00)$ and $(11\bar{2}0)$ surfaces	47
		3.4.4 STM images of non-polar wurtzite surfaces	49

4 Experimental setup   **53**
	4.1	The UHV chamber system	53
	4.2	Tip preparation	54
	4.3	Sample preparation	55
	4.4	Cleavage of the sample and tip approach	55

5 Structural and electronic properties of diluted GaAsN   **57**
	5.1	Sample structure and growth	58
	5.2	Nitrogen incorporation into GaAs	59
	5.3	Identification of single nitrogen atoms	59
	5.4	Nitrogen versus arsenic vacancy in the GaAs(110) surface	61
	5.5	Local density of states	63
		5.5.1 STS results	63
		5.5.2 Comparison with the BAC model calculations	65
		5.5.3 Green's function approach	66

6 InAs(N)/GaAs(N) quantum dots   **71**
	6.1	InAs quantum dots within a nitrogen containing GaAs matrix	71
		6.1.1 Sample structure and growth	72
		6.1.2 Structural changes induced by the nitrogen	73
		6.1.3 Stoichiometry determination	74
		6.1.4 Optical characterization	76
		6.1.5 Discussion	77
	6.2	InAsN/GaAs quantum dots	77

		6.2.1	Sample structure and growth	78

		6.2.1	Sample structure and growth	78
		6.2.2	The reference bilayer	79
		6.2.3	The $InAs_{0.91}N_{0.09}$/GaAs bilayer	81
		6.2.4	Optical characterization	84
		6.2.5	Phase separation between indium and nitrogen ..	85
7	**Non-polar GaN surfaces**			**89**
	7.1		The GaN($1\bar{1}00$) cleavage surface	91
	7.2		STS investigations of the non-polar GaN$(1\bar{1}00)$ cleavage surface	93
		7.2.1	Density of states	93
		7.2.2	Intrinsic surface states dispersion	96
		7.2.3	Discussion	97
	7.3		Dislocations in GaN	98
		7.3.1	Density of dislocations	99
		7.3.2	Determination of the line direction from STM images	100
		7.3.3	Perfect dislocation	101
		7.3.4	Dissociated dislocation	102
		7.3.5	Charge of dislocations	104
		7.3.6	Discussion	106
	7.4		Doping modulation in GaN	106
		7.4.1	STM investigations	107
		7.4.2	Secondary ion mass spectroscopy investigations	109
		7.4.3	Origin of the observed modulation	110
		7.4.4	Theoretical simulation of the topography modulation	112
8	**Conclusion**			**115**
	Bibliography			**119**
	Acknowledgments			**135**

Chapter 1

Introduction

Solid materials are used because of their deformation resistance. Over the centuries the originally available natural minerals were completed or replaced by synthetically created materials like metal alloys, glasses, and ceramics. The industrial development led to the request for better and variable materials. The invention of the transistor (1947) by J. Bardeeen [1] laid the foundation for the semiconductor industry. Nowadays, it can be estimated that more than 1000 transistors are produced, per day and per person on the planet [2]. Devices like the MOS transistor (1965), the light-emitting diode (LED) (1960), and the microprocessor (1971) are the necessary building blocks for the consumer electronic items that are now useful to everyone. It was the miniaturization of products made possible by the semiconductor devices, which allowed the personalization of electronics and a revolution in living style.

III-V compound semiconductors provide the material basis for many technologies and applications due to the ability to synthesize materials with a tunable energy band gap by changing the composition of their alloys [3–5]. In this way, e.g. the wavelength of the light emission can be adjusted. The emission wavelength of III-V based semiconductor lasers covers almost the entire spectral range from ultraviolet to the mid infrared [2, 6, 7]. Fig. 1.1 demonstrates which material can be used in order to reach a requested wavelength. In this figure the fundamental band gap dependence on the lattice constant is demonstrated. The different material systems can be classified by the spectral range they cover: *nitride-based* from ultraviolet to blue-green, *phosphide-based* from yellow-green to near-infrared, *arsenide/phosphide-based* for infrared, and *antimonide-based* for far-infrared.

A wide range of III-V heterostructures has been investigated for their properties, although only a few are commonly used. However, GaAs is the most technologically important and the most studied compound semiconductor material [6]. InGaAs

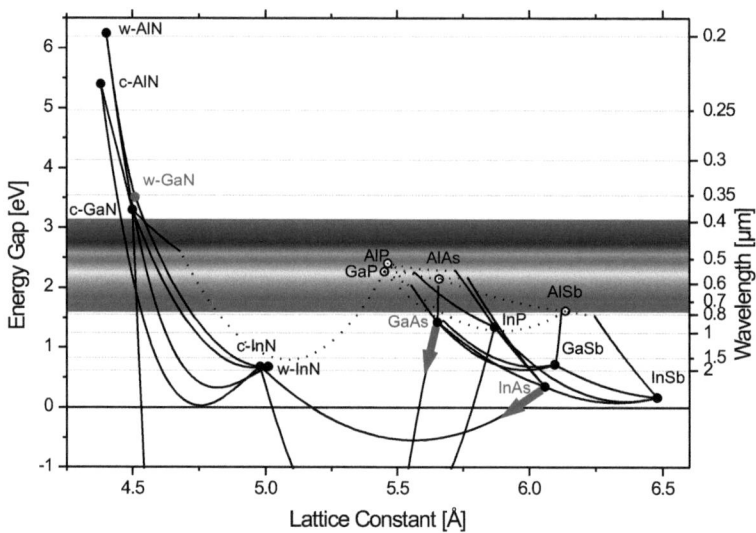

Figure 1.1: Fundamental band gap of various III-V semiconductors as a function of their lattice constant [8]. Solid lines represent a direct band gap, while dotted lines indicate an indirect gap. The rainbow bar represents the visible light range. GaN, GaAsN, and InAsN semiconductors are studied in this thesis and therefore are marked in red. The red arrows indicate the band bowing of GaAsN and InAsN at low nitrogen concentration.

alloys on GaAs substrates are a key component in the active regions of high-speed electronic devices [9], storage devices [10, 11], infrared lasers [12–15], and long-wavelength quantum cascade lasers [16]. They are in particular used as active region in long-wavelength lasers for optical communication, at emission wavelengths of 1.3 μm and 1.55 μm, where the widely used single-mode optical fibers have a chromatic dispersion minimum and an attenuation minimum, respectively [17].

In the second half of the 1990s an expansion of research on nitrogen-containing semiconductors occurred. This interest has continued unabated to the present. First the diluted nitride heterostructures (nitrogen concentration below 5%) were introduced as possible candidates to expand the emission wavelength of the GaAs-based devices towards 1.3 μm and even 1.55 μm [18, 19]. The exceptional red-shift of the emission wavelength of diluted GaAsN/GaAs quantum wells (QWs) and quantum dots (QD)s is due to the extremely strong band gap reduction caused by already low nitrogen incorporation [18,19], as indicated in Fig. 1.1 by a red arrow.

Further, pure GaN and related compounds attracted a wide technological importance despite their high defect densities [20]. Wide band gap semiconductor materials extend the field of semiconductor applications to the limits where classical semiconductors such as Si and GaAs fail [21]. Group-III nitride semiconductors are outstanding candidates for a broad range of optoelectronic device applications in the green-to-ultraviolet spectral range [22,23]. Moreover, they can operate at higher temperatures due to the high thermal conductivity [2]. GaInN QWs represent a key constituent in the active regions of blue diode lasers and LEDs [24–30], GaN/AlN QDs emit in the ultra-violet spectral region [31–36], and InN/GaN QDs can in principle extend the spectral range of nitride QDs to the mid green [37]. Not only from the technological, but also from the academical point of view, the nitrides are fascinating semiconductors, since GaN and its alloys exhibit very unusual physical properties.

In this thesis the structural as well as the electronic properties of both diluted Ga(In)As nitrides and pure GaN semiconductors are investigated. Although in many ways these two classes of nitrogen-containing III-V materials are not closely related, they demonstrate the common attribute to gain a wide attention in a very short period of time [38]. The structure of this work is the following. Chapter 2 contains the theoretical background of structural and electronic properties of nitrogen containing III-V compound semiconductors. Scanning tunneling microscopy, the experimental method used in this work, is described in chapter 3, whereas the experimental setup is reviewed in chapter 4. In chapter 5 the structural and spectroscopic results of GaAsN are presented, followed in chapter 6 by the discussion of the structural properties of InAs(N)/GaAs(N) QDs. Chapter 7 describes the structural and electronic properties of GaN and the non-polar GaN($1\bar{1}00$) surface. Finally, chapter 8 summarizes the results of this work.

Chapter 2

Nitrogen containing III-V compound semiconductors

2.1 Crystal structure of group-III arsenides versus group-III nitrides

Group-III arsenides crystallize in the cubic zincblende structure. For group-III nitrides the wurtzite structure is typical, but the cubic zincblende structure is as well possible. In both crystal structures each group-III atom has four group-V atoms as nearest neighbors and vice versa (see Fig. 2.1). The zincblende crystal is build from two interpenetrating face-centered cubic (fcc) sublattices spaced 1/4 along the body diagonal and belongs to the $F\bar{4}3m(T_d)$ space group [39].

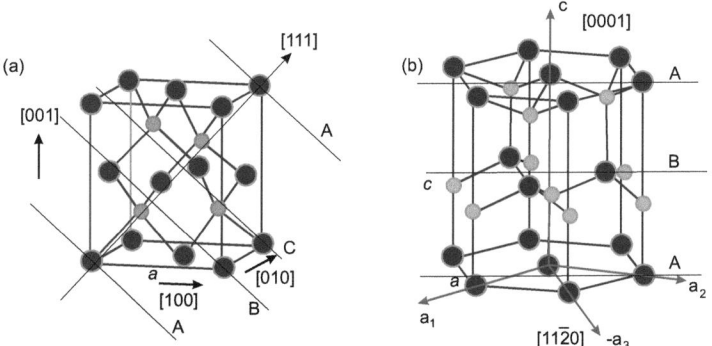

Figure 2.1: *(a) Cubic zincblende structure with the lattice constant a, and (b) wurtzite structure with the lattice constants a and c.*

The wurtzite structure is a combination of two hexagonal close-packed (hcp) sublattices which are displaced by $5/8c$ relative to each other along the [0001] direction. It can be described simply with reference to the stacking sequence. The lattice is formed by the stacking of close-packed planes, as in the cubic zincblende structure along the <111> directions, but the stacking sequence is now $\{ABABAB\ldots\}$, instead of $\{ABCABCA\ldots\}$ for the cubic case [see Fig. 2.1(b)]. The crystal belongs to the $P6_3mv(C_{6v})$ space group [39]. The close-packed plane perpendicular to the c axis is called the basal plane. The ideal ratio $u = c/a$ of the wurtzite structure is theoretically 1.633, but for real crystals it varies depending on the electronic structure of the atoms. If Miller indices of three numbers based on axes a_1, a_2, and c are used to define planes and directions in the wurtzite structure, it is found that crystallographically equivalent planes can have combinations of different numbers. Indexing in hexagonal crystals is therefore usually based on Miller-Bravais indices, which are referred to the four axes a_1, a_2, a_3, and c [see Fig. 2.1(b)]. When the reciprocal intercepts of a plane on all four axes are found and reduced to the smallest integers, the indices are of the type (h, k, i, l) and the first three indices are related by $h + k + i = 0$.

2.2 Band bowing and band-anticrossing model in III-V nitrides

2.2.1 Vegard law

The lattice constants of ternary alloys are usually described by the Vegard law [40], stating that the lattice constant of a ternary alloy can be expressed as a linear combination of the lattice constants of the two composing binary compounds. This approximate empirical rule predicts for the lattice constant of an alloy

$$a(A_xB_{1-x}Z) = x \cdot a(AZ) + (1-x) \cdot a(BZ),$$

where AZ and BZ are the binary compounds, ABZ the ternary alloy and x the concentration of A with $x \in [0,1]$. Whereas the Vegard law shows a linear correlation, the dependence of the energy gap on the alloy composition can be expressed in most cases by a quadratic form:

$$E_g(A_xB_{1-x}Z) = x \cdot E_g(AZ) + (1-x) \cdot E_g(BZ) - x \cdot (1-x) \cdot C(A_xB_{1-x}Z),$$

where the so-called bowing parameter C represents the deviation from a linear interpolation (virtual-crystal approximation) between the two binary compounds AZ and

BZ. Ideally C is independent of the concetration x. The bowing parameter for III-V alloys is typically positive, i.e., the alloy band gap is smaller than the linear interpolation result (see Fig. 1.1). The physical origin of the band gap bowing can be traced to disorder effects created by the presence of different cations or anions [41]. The bowing concept has been generalized to include quadratic terms in the alloy-composition dependence of several other band parameters, which in some cases may be attributable to specific physical mechanisms but in others simply represent empirical fits to the experimental data.

2.2.2 Diluted GaAsN

Diluted GaAsN with a nitrogen concentration of up to 5% has received high attention for both technological and fundamental reasons [42]. Since GaAsN is formed from GaAs and GaN, both having a direct band gap of 1.43 eV and 3.4 eV, respectively, one could expect diluted GaAsN to cover the visible spectrum which lies between near infrared and ultraviolet. However, this is not the case (see Fig. 1.1). The GaAsN band gap decreases by approximately 180 meV per 1% nitrogen concentration increase [43,44]. This leads to the conclusion of an unexpectedly large bowing of the band gap, which could even result in a negative band gap and thus metallic behavior at large nitrogen concentrations. The bowing parameter C is found to be about $(20-25)$ eV for $x < 1\%$ and $(15-20)$ eV for $x > 1\%$ [45]. This band gap bowing is attributed to the interaction between a specific narrow band of localized nitride states and the conduction band (CB) of the GaAs host crystal [46].

The large lattice mismatch between cubic GaAs and cubic GaN (about 20%, see Tab. 2.1) limits the growth of GaAsN, so that only nitrogen concentrations of up to 5% [38] can be achieved in GaAsN compounds.

Moreover, the substitution of arsenic atoms by nitrogen atoms results in strong band modifications since there are significant differences between these two elements in covalent radii [54,55] (0.70 Å for nitrogen and 1.19 Å for arsenic) and electronegativity [56] (see Tab. 2.2), and furthermore the Ga-N bond is more than twice as stiff as the Ga-As bond [57].

It is known that a bound impurity state may exist when the impurity and host atoms have sufficiently different properties. Due to its electronegative character, nitrogen in small quantities introduces an acceptor level in GaAs close to the CB edge. Using the band anticrossing (BAC) model, Shan et al. [46] proposed that these acceptor levels are very localized in real space (thus very delocalized in the reciprocal space according to the Heisenberg uncertainty relation) and therefore interact with

Parameters	arsenides			phosphides		
at 300 K	AlAs	GaAs	InAs	AlP	GaP	InP
E_G (eV)	2.16	1.43	0.36	2.45	2.26	1.35
a (Å)	5.66	5.65	6.06	5.47	5.45	5.87
f_i	0.274	0.310	0.357	0.307	0.327	0.421
c_{11} (gPa)	1250	1221	833	1330	1405	1011
c_{12} (gPa)	534	566	453	630	620	561
c_{44} (gPa)	542	600	396	615	703	456
d_0 (Å)	2.44	2.45	2.62	2.37	2.36	2.54
E_B (eV)	1.89	1.63	1.55	2.13	1.78	1.74
Parameters	cubic nitrides			wurtzite nitrides		
at 300 K	AlN	GaN	InN	AlN	GaN	InN
E_G (eV)	5.20	3.23	1.92	6.20	3.42	0.67
a (Å)	4.38	4.52	4.98	3.11	3.19	3.55
c (Å)				4.98	5.19	5.76
f_i	0.449	0.500	0.578	0.449	0.500	0.578
c_{11} (gPa)	315	291	187	410	373	190
c_{12} (gPa)	150	148	125	140	141	104
c_{13} (gPa)				100	80	121
c_{33} (gPa)				390	387	182
c_{44} (gPa)	185	158	86	120	94	10
c_{66} (gPa)				135	116	43
d_0 (Å)	1.90	1.95	2.16	1.90	1.95	2.16
E_B (eV)	2.88	2.32	1.93	2.88	2.32	1.93

Table 2.1: Band gaps E_G [6,47,48], lattice constants a and c [6,48], ionicities f_i [6,49–51], c_{ij} stiffness constants [6,48], bond lengths d_0 (Å) [52], and cohesive energies per bond E_B [53] of arsenide, phosphide, and nitride compound semiconductors.

the delocalized states of the GaAs CB edge, leading to two possible solutions:

$$E_\pm(k) = \frac{1}{2}\left([E^{\mathrm{C}}(k) + E^{\mathrm{N}}] \pm \sqrt{[E^{\mathrm{C}}(k) - E^{\mathrm{N}}]^2 + 4V^2 x}\right), \qquad (2.1)$$

group–III element	covalent radius (Å)	electronegativity
B	0.84	2.04
Al	1.21	1.61
Ga	1.22	1.81
In	1.42	1.78
group–V element	covalent radius (Å)	electronegativity
N	0.71	3.05
P	1.07	2.19
As	1.19	2.18
Sb	1.39	2.09
Bi	1.48	2.02

Table 2.2: *Covalent radii [55] and electronegativities [56] of the different group-III and group-V elements.*

where $E^C(k)$ is the CB dispersion of the unperturbed nitride free GaAs, with $E^C(0) = 1.45$ eV, E^N is the position of the nitrogen iso-electronic impurity level, V is the interaction potential between the two bands, and x is the nitrogen fraction. The resulting dispersion relation for the two coupled conduction bands in GaAs$_{0.988}$N$_{0.012}$, together with the characteristic anticrossing, is shown in Fig. 2.2(a) for k values between $-0.15/$Å and $0.15/$Å. For higher k values the conduction and the E_+ band overlap.

In Fig. 2.2(b) the fundamental band gap between the valence band (VB) maximum and the E_- CB minimum is plotted as a function of nitrogen fraction x for GaAs$_{1-x}$N$_x$ at 300 K for different models (BAC, band bowing), as well as for experimental data [38]. A curve with a constant bowing parameter of 18 eV (dotted), and also a curve with a variable bowing parameter of $(20.4 - 100x)$ eV (dashed) are displayed for comparison. It can be seen that the BAC model predicts substantially higher energy gaps for nitrogen fractions exceeding 1.5%. The available experimental data (points) compiled in Ref. [58] clearly show a much better agreement with the BAC parameterization than with either of the two curves simply based on empirical bowing parameters.

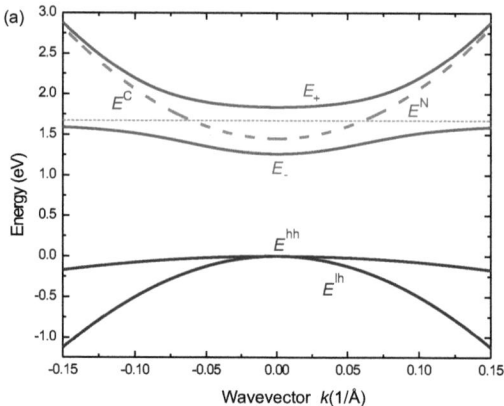

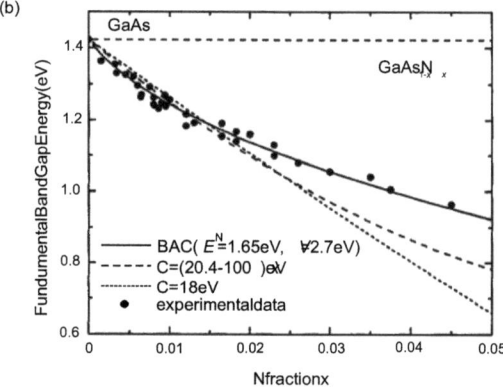

Figure 2.2: (a) Valence and conduction-band dispersion relations for GaAs$_{0.988}$N$_{0.012}$ at 300 K from the BAC model (solid curves). For comparison, the unperturbed GaAs conduction band (dashed parabola) and the position of the nitrogen level (dashed line) are shown. The nitrogen effects on the valence bands are completely neglected. E^N=1.65 eV and V=2.7 eV are used. (b) Energy of the fundamental band gap transition in GaAsN as a function of nitrogen fraction x from the BAC model (solid curve), using a variable bowing parameter (dashed curve), and using a constant bowing parameter (dotted curve) [38]. For comparison, the experimental data from [58] are plotted (circles).

The BAC model provides a useful and reliable basis for the description of a variety of diluted nitride material properties, such as the fundamental energy gap (governed by the transition from the top of the VB to E_-), the temperature shift of the gap, the electron effective mass, and the characteristics of the upper band E_+ (within the theory of Lindsay et al. [59]). The BAC model does not consider anything else than a single nitrogen level on a substitutional lattice site (or a narrow impurity band formed from such levels). It neglects not only mixing with the L and X valleys, but also a more complex behavior of nitrogen in the semiconductor matrix, such as the formation of nitrogen pairs and clusters [38].

2.2.3 Diluted GaInAsN

Most of the technological interest in diluted nitrides has so far focused on GaInAsN heterostructures grown on GaAs substrates [60], since it is promising for long wavelength telecommunication lasers and solar cells. It was predicted that in contrast to ternary GaInAs/GaAs QWs (which would have too much compressive strain to reach the desired wavelength range), the addition of nitrogen narrows the band gap for lower indium concentrations, accompanied by the introduction of tensile strain. Thus, the compressive strain can be compensated by the tensile one, in such a way that the overall active region can be successfully lattice-matched to the GaAs substrate.

The applicability of the BAC parameterization to GaInAsN is well established [46, 61–63]. Studies of InGaAsN with low indium concentration [46, 62] on the order of 10% or less report no significant differences from GaAsN apart from the generally lower energy gap due to the indium content. However, a smaller band gap reduction for GaInAsN than in GaAsN is theoretically expected [64, 65] due to ordering of the nitrogen atoms in the GaInAs matrix [66, 67].

2.3 Low-dimensional semiconductor structures

A semiconductor structure has a reduced dimensionality when the motion of at least one type of charge carriers is confined in at least one direction within a spatial size in the order of their de-Broglie wavelength [68]

$$\lambda_\mathrm{B} = \frac{h}{(2m_\mathrm{e}) \cdot k_\mathrm{B} T},$$

being e.g. $\lambda_\mathrm{B} \approx 7.6$ nm at 300 K. The carrier momentum in the respective direction gets quantized and its energy spectrum is given by the discrete solutions of the Schrödinger equation, the eigenenergies. As a consequence the carrier has a non-vanishing minimum kinetic energy, the quantum confinement energy. For confinement in one, two, or three dimensions the expressions QW, quantum wire, and QD have

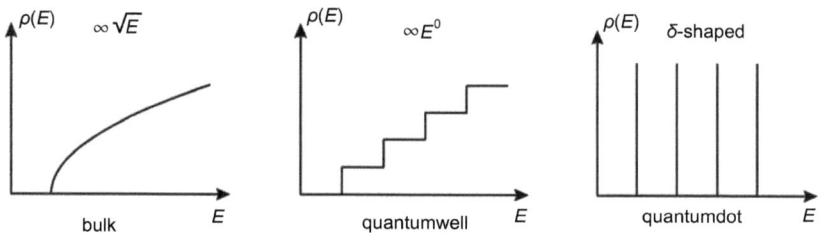

Figure 2.3: Density of states for carriers in three, two, and one dimensions.

been established, respectively.

As an important consequence of the reduced dimensionality the density of states $\rho(E)$ of the quantum confined carrier changes according to the dimensionality of the structure. For three-dimensional bulk the density of states is given by

$$\rho_{3D}(E) \sim \frac{1}{2\pi^2}\left(\frac{2m^*}{\hbar^2}\right)^{3/2}\sqrt{E}. \tag{2.2}$$

For a two-dimensional QW the density of states is given by

$$\rho_{2D}(E) \sim \frac{1}{L_z\pi}\frac{m^*}{\hbar^2}\sum_{i=0}^{n}\theta(E-E_i). \tag{2.3}$$

For a zero-dimensional QD the density of states is given by

$$\rho_{0D}(E) \sim \frac{1}{L_xL_yL_z\pi}\left(\frac{m^*}{\pi^2\hbar^2}\right)^{1/2}\sum_{i=0}^{n}\delta(E-E_i). \tag{2.4}$$

L_x, L_y, L_z are the sizes of the structure in x, y, and z direction, E_i the i-th eigenenergy of the carrier, m^* the effective carrier mass, θ the unit step function [$\theta(x,\ x<0)=0$], and [$\theta(x,\ x\geq 0)=1$], δ the delta function, and $\hbar$ the reduced Planck constant. The density of states of these differently confined systems are schematically shown in Fig. 2.3.

2.3.1 Bound particles

The simplest example of a QW is shown in Fig. 2.4. The electron has zero potential energy in the region $0<x<a$, with a being the width of the well, and infinitely high potential barriers prevent it from scattering beyond this region. The solution of the Schrödinger equation provides the wave functions and energies:

$$\Phi(x) = A_n \sin\frac{n\pi x}{a}, \tag{2.5}$$

2.3 Low-dimensional semiconductor structures

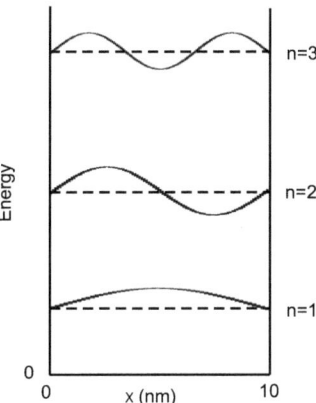

Figure 2.4: *Infinitely deep square well of 10 nm width along x, showing the first three energy levels and wave function [68].*

$$E_n(k_n) = \frac{\hbar^2 k_n^2}{2m^*} = \frac{\hbar^2 \pi^2 n^2}{2m^* a^2}. \tag{2.6}$$

The integer $n = 1, 2, 3, \ldots$ is a quantum number that labels the states. Quantisation – the restriction on the allowed energies – has come from the boundary conditions that constrain the motion of the particle. Particles that can move freely over all space have a continuous range of energy levels, while those that are confined to a region of space have discrete levels. The wave functions in the QW have an important symmetry property. Those with odd n are symmetric functions respective to the center of the well whereas those with even n are antisymmetric functions.

2.3.2 Formation of low-dimensional structures – Molecular beam epitaxy

Semiconductor heterostructure must have interfaces of high quality and their composition needs to be changeable, preferably from one monolayer (ML) to another. Molecular beam epitaxy (MBE) is the most widespread method for the growth of diluted nitrides and is in principle a very simple method. It is a powerful and flexible technology for research and development as well as for mass production of nanoelectronic structures based on a huge variety of materials [69].

The elements that compose the heterostructure are usually vaporized in Knudsen cells. Molecules that emerge from these cells form a molecular beam, which moves statistically towards the sample substrate. Growth rates and the flux of each element can be controlled through the temperature of each source. Because of the vacuum deposition, MBE growth is performed under conditions far from thermodynamic equilibrium and is governed mainly by the kinetics of the surface processes occurring when the impinging beams react with the sample surface.

MBE is a rather slow process, the amount of material deposition is at about 1 MLs^{-1} or 1 μmh^{-1}. Furthermore only small-sized wafers can be covered. The advantage of MBE include highly abrupt junctions between different materials, good control over the thickness of layers, and reasonable reproducibility. Moreover, MBE can be controlled in situ by surface sensitive diagnostic methods such as reflection high-energy diffraction. Obvious disadvantages include the cost and the difficulties of applying the process for production. Additional details of MBE can be found in references [69, 70].

2.3.3 Strained layers

When two materials have different lattice constants, such as InAs and GaAs, three different results are possible, when InAs is epitaxially grown on GaAs, as shown in Fig. 2.5. In equilibrium InAs has a larger lattice constant [Fig.2.5(a)]. It is assumed that the GaAs layer is so thick that it cannot be distorted significantly. If the InAs layer is thin, it can get compressively strained to fit the GaAs in the interface, as shown in Fig. 2.5(b). Thus, the lattice constant of InAs in the plane is reduced to that of GaAs, while the usual elastic response causes it to get extended along the z direction. This results in a tetragonal distortion and a considerable increase of elastic energy. Thus, the stress is enormous in comparison with the macroscopic scale.

The elastic energy of a grown strained layer is roughly proportional to its thickness. If InAs is deposited on GaAs(001) in the range of 1.5 to 1.7 ML the InAs forms a flat but strained layer, as discussed above and schematically shown in Fig. 2.5(b) [71]. This is the so-called wetting layer. If the thickness of the InAs layer is increased, two different cases can occur, as shown in Fig. 2.5(c) and (d). At InAs layer coverages higher than 1.7 ML three-dimensional islands start to form due to the large lattice-mismatch of InAs respective to GaAs amounting to about 7% [71], as shown in Fig. 2.5(c). Thereby the lateral positions of the atoms within the islands are slightly shifted in respect to those of the wetting layer atoms, but the crystal lattice and the symmetry are continued [72].

With proceeding deposition, the size of the QDs increases, until the strain within these structures gets too large. If the InAs coverage is than further increased, the stress can be built up only by the relaxation in order to reach the natural InAs lattice

2.4 Surface segregation

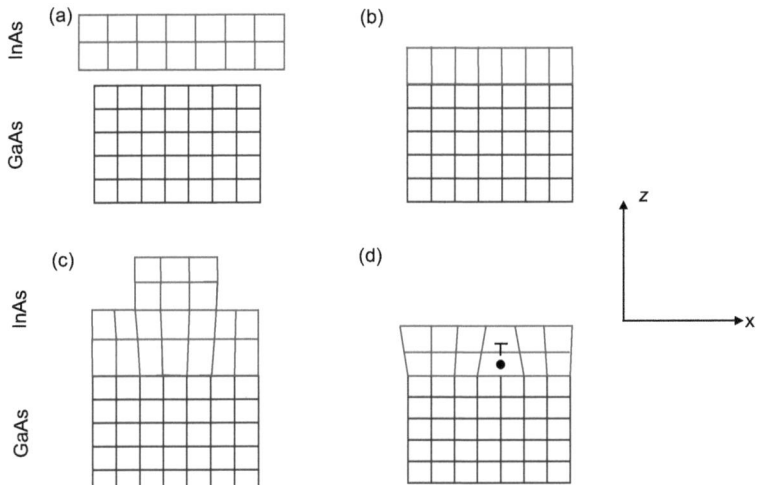

Figure 2.5: *Growth of InAs on GaAs. (a) Separate layers at equilibrium. (b) Thin layer of InAs on GaAs. The InAs gets strained to fit the lattice constant of GaAs in the plane of the heterojunction (Frank-van der Merwe growth). (c) QD formation in the Stranski-Krastanow growth. (d) Thicker layer, where the strain has relaxed, and a misfit dislocation at the heterointerface occurs, as indicated by a T.*

constant, so that perfect matching of the atoms at the heterojunction is no longer possible. Instead the difference in lattice constants is balanced by the appearance of misfit dislocations, as shown in Fig. 2.5(d). Bonds get broken when the dislocations form, and this requires an energy proportional to the sample area. Therefore a critical thickness exists beyond which it is energetically more favorable for dislocations to appear [68].

2.4 Surface segregation

2.4.1 Theory of segregation

The term surface segregation refers to the enrichment of solute atoms of a constituent at an external surface or at an internal interface of a material, relative to bulk composition [73]. In contrast to the diffusion, which result is a gradual mixing of material atoms, the result of the segregation process is a demixing of material atoms. The driving force behind the segregation is on the one hand the minimization of the surface energy (thermodynamic approach) and on the other hand the strain relaxation (kinetic approach). Since the surface energy and the bulk energy are different, solute atoms segregate to the surface in accordance with the statistics of thermodynamics in order to minimize the overall free energy of the system [74]. But especially during epitaxial growth the segregating material atoms do not have sufficient time to reach their equilibrium level as defined by the thermodynamical theories. The kinetics of segregation becomes a limiting factor. This happens further, if the segregation occurs as a result of solute atoms coupling to vacancies which are moving to boundary surfaces or interfaces during application of stress [75]. In a solid, segregation can occur in particular at dislocations, grain boundaries, or heterostructures interfaces of semiconductors with different lattice constants.

2.4.2 Intermixing at InAs/GaAs and GaAs/InAs interfaces

At MBE and MOCVD growth temperatures, atomic diffusion is not possible in the bulk [76]. The structure geometry is determined at the growth surface, where atomic movements at this surface are easier, leading to exchanges between substrate atoms still uncovered and impinging atoms. For example, during the formation of AlAs/GaAs interfaces, Al-Ga exchanges drive Ga atoms to the top of the growing overlayer, and for Si/GaAs, Ge/GaAs, and Si/GaP both substrate atoms are involved [77].

The exchange is not necessarily observed for both A/B and B/A interfaces. For InAs/GaAs interfaces, indium atoms stay on GaAs to form InAs with neighboring arsenic atoms. For GaAs/InAs interfaces, in contrast gallium atoms are incorporated underneath an upper InAs ML, which floats on the growing GaAs layer; however, this reaction is not complete, so that indium gradually dissolves in the first 20 Å of this overlayer [76]. Similar results have been reported for the growth of AlAs on GaAs (only gallium segregation) and GaSb on AlSb (only aluminum segregation) [76].

Indium and gallium atoms have different bonding energies to arsenic (In: 1.55 eV, Ga: 1.63 eV) and covalent radii (In: 1.42 Å, Ga: 1.22 Å) (see Tab. 2.1 and Tab. 2.2), so that the driving force for segregation may originate from a bond-strength or a steric effect: the more weakly-bonded or biggest atom streams out to the surface.

The elastic strain may also play an important role (the biggest atom comes to the surface). Indeed, the bond length relation is similar: the In-As bond (2.62 Å) is larger than the Ga-As bond (2.45 Å). These effects are also modified by bond distortions due to the strain and to the surface reconstruction, both variable during the heterojunction growth [76].

2.5 Dislocations and stacking faults in the wurtzite structure

Material quality has always been a central issue for the technological development of semiconductors. Unfortunately, no well-suited lattice- and thermal-matched substrates are available for epitaxial GaN growth, leading to a huge amount of strain. The reduction of strain can occur by inserting dislocation [78].

2.5.1 Defects in crystals

All actual crystals contain imperfections such as point, line, surface, or volume defects. They disturb locally the regular arrangement of the atomic lattice. Their presence may significantly modify the properties of the respective solid. Line defects, called dislocations, are described by a dislocation line in the direction $\underline{u}$ and the Burgers vector $\underline{b}$ [79]. A Burgers circuit is any atom-to-atom path taken in a crystal containing dislocations which forms a closed loop. Such a path is marked in Fig. 2.6(a) for an

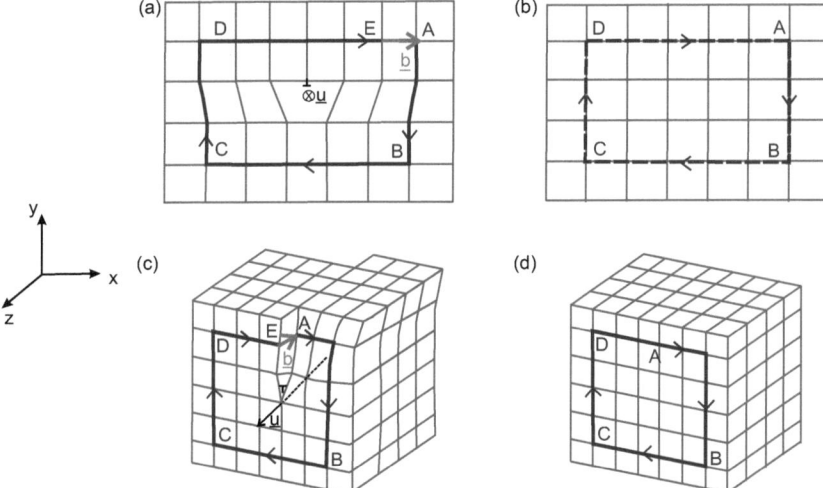

Figure 2.6: (a) Burgers circuit around an edge dislocation, (b) the same circuit in a perfect crystal. Vector $\underline{b}$ is the Burgers vector and $\underline{u}$ marks the dislocation direction. The dislocation line is along the z direction perpendicular to the Burgers vector $\underline{b}$. (c) Burgers circuit around a screw dislocation, (d) the same circuit in a perfect crystal. The dislocation line $\underline{u}$ is along the z direction and parallel to the Burgers vector $\underline{b}$.

edge dislocation and (b) for a dislocation-free crystal by $ABCDE$. Vector $\underline{b}$ is the Burgers vector. The dislocation line is along the z direction, normal to the Burgers vector.

If the same atom-to-atom sequence is followed in a dislocation-free and a dislocation-containing crystal, then the latter circuit will not close, if this circuit encloses one or more dislocations. The vector required to complete the circuit is called the Burgers vector. Is the Burgers vector normal to the line of the dislocation, than this dislocation is called an edge dislocation [Fig. 2.6(a)] [79]. Is the Burgers vector parallel to the line of the dislocation, than the dislocation is called a screw dislocation [Fig. 2.6(c)] [79]. In the most general case the dislocation line lies at an arbitrary angle to its Burgers vector, and the dislocation line has a mixed edge and screw character [79]. However, the Burgers vector of a single dislocation has fixed length and direction, and is independent of the position and orientation of the dislocation line. A dislocation whose Burgers vector is a lattice translation vector is called a perfect or unit dislocation, otherwise this dislocation is called partial or imperfect dislocation [79]. In the latter case the regular sequence of the crystal (e.g. $ABCABC$ for the cubic case) has been interrupted, indicating a planar defect, called stacking fault, which is shown in Fig. 2.7.

Dislocations do not begin or end anywhere in a crystal. They may form closed loops inside a crystal or begin and end on edges, interfaces or grain boundaries of the crystal, are introduced at interfaces or branch into other dislocations. The latter gives two (or more) dislocation lines with new Burgers vectors. The sum of these Burgers vectors is always equal to the Burgers vector of the initial dislocation [79].

2.5.2 Dislocations and stacking faults in the wurtzite structure

In the following, dislocations in the hcp structure with the line direction along the c axis will be discussed in detail according to the Ref. [79]. Their Burgers vectors can be described by the bi-pyramid shown in Fig. 2.8. The most important dislocations with their Burgers vectors are listed in Table 2.3. They are discussed in the following:
(a) Six perfect edge dislocations with Burgers vectors in the basal plane along the sides of the triangular base ABC of the pyramid. They are $\underline{AB}$, $\underline{BC}$, $\underline{CA}$, $\underline{BA}$, $\underline{CB}$ and $\underline{AC}$. The corresponding vectors are marked in Fig. 2.8 in yellow.
(b) Two perfect screw dislocations with Burgers vectors perpendicular to the basal plane, represented by the vectors $\underline{ST}$ and $\underline{TS}$, marked in blue.
(c) Twelve perfect mixed dislocations whose Burgers vectors are represented e.g. by

2.5 Dislocations and stacking faults in the wurtzite structure

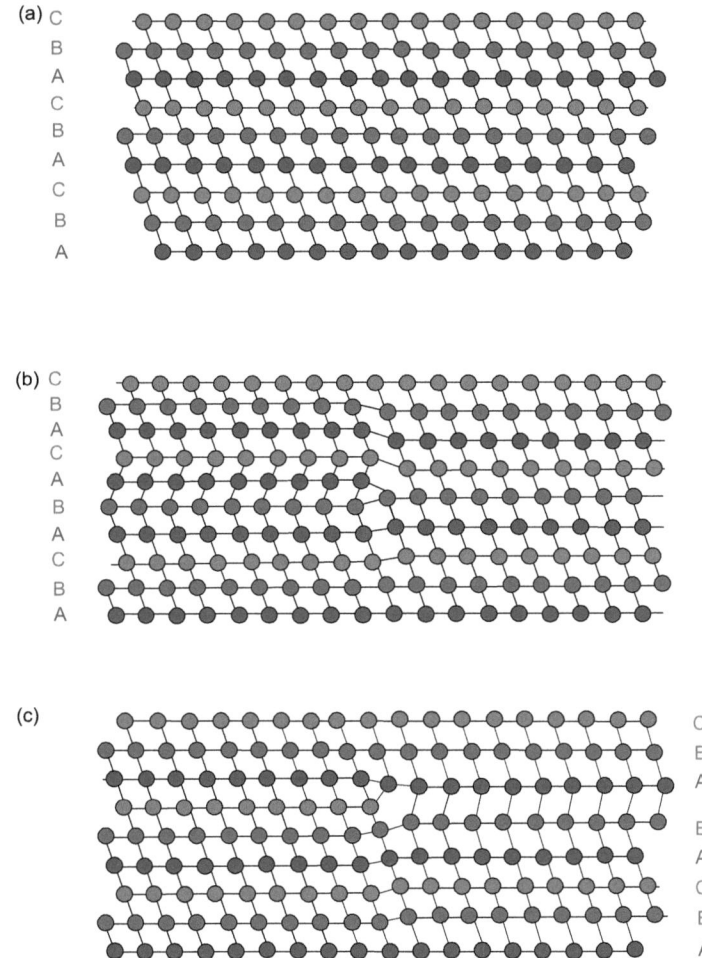

Figure 2.7: *Stacking faults in cubic crystals, corresponding to partial dislocations: (a) Perfect stacking sequences ABCABCA.... (b) An interstitial plane is filled. The cubic stacking sequence is faulty, being now {ABCABACABC...}. (c) A vacancy plane is created, resulting in a stacking fault because the cubic stacking sequence has been changed to the faulty sequence {ABCABABC...}.*

symbols such as $\underline{AD}$, being $\underline{AD} = \underline{AB} + \underline{TS}$. The corresponding vector is marked in green.
(d) Four imperfect dislocations of type I with Burgers vectors perpendicular to the

Typical Burgers vectors	Notation	Description
$\underline{AB}$, $\underline{AC}$, $\underline{BC}$	$\frac{1}{3}\langle 11\bar{2}0\rangle$	Perfect edge dislocations
$\underline{ST}$	$\langle 0001\rangle$	Perfect screw dislocations
$\underline{AB}+\underline{TS}$	$\frac{1}{3}\langle 11\bar{2}3\rangle$	Perfect mixed dislocations
$\underline{\sigma S}$, $\underline{\sigma T}$	$\frac{1}{2}\langle 0001\rangle$	Imperfect dislocations type I
$\underline{A\sigma}$, $\underline{B\sigma}$, $\underline{C\sigma}$	$\frac{1}{3}\langle 10\bar{1}0\rangle$	Shockley partials (imperfect dislocations type II)
$\underline{AS}$, $\underline{BS}$	$\frac{1}{6}\langle \bar{2}203\rangle$	Combination of the types I and II (type III)

Table 2.3: *Dislocation types with [0001] line direction present in wurtzite GaN structures [79].*

basal plane, namely, $\underline{\sigma S}$, $\underline{\sigma T}$, $\underline{S\sigma}$, and $\underline{T\sigma}$, marked in orange.

(e) Six imperfect basal dislocations of type II or so-called Shockley partial dislocations with Burgers vectors $\underline{A\sigma}$, $\underline{B\sigma}$, $\underline{C\sigma}$, $\underline{\sigma A}$, $\underline{\sigma B}$, and $\underline{\sigma C}$, marked in brown.

(f) Twelve imperfect dislocations which are a combination of the latter two types given by $\underline{AS}$, $\underline{BS}$, etc, marked in violet. Although these vectors represent a displacement from one atomic site to another the associated dislocations are not perfect. This is because the sites do not have identical surroundings and the vectors are not translations of the lattice [79].

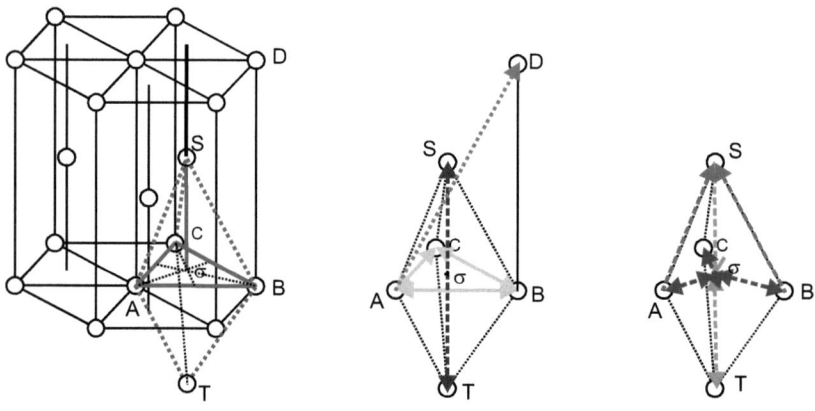

Figure 2.8: *Burgers vectors in the hexagonal close-packed lattice, see text.*

Imperfect dislocations change the regular stacking sequence $\{ABABAB\ldots\}$ of the layers. Such a breaking of the stacking sequence is a planar defect, called stacking fault. Three different types of stacking faults can be classified, depending on their Burgers vector: $\{ABABCABAB\ldots\}$, $\{ABABCBCBC\ldots\}$, or $\{ABABCACAC\ldots\}$ (see Fig. 2.9). A stacking fault may cause a step on the $(1\bar{1}00)$ surface between two parts of the crystal, which are separated by the plane of the stacking fault. In the case of the sequence $\{ABABCABAB\ldots\}$ no height difference occurs on the $(1\bar{1}00)$ surface. The sequences $\{ABABCBCBC\ldots\}$ and $\{ABABCACAC\ldots\}$ yield a step with 1/3 of the monoatomic step height on the $(1\bar{1}00)$ surface.

2.5.3 Movement of dislocations

The atomic displacement caused by a dislocation produces a stress field in the crystal around the dislocation [80]. After their nucleation dislocations can move through the crystal in order to minimize their elastic energy [79]. There are two basic types of dislocation movement, glide or conservative motion [Fig. 2.10(a)], in which the dislocation moves in the plane which contains both its line and the Burgers vector, and climb or non-conservative motion [Fig. 2.10(b)] in which the dislocation moves out of the glide plane normal to the Burgers vector [79].

At low temperatures where diffusion is not dominant, and in the absence of a non-equilibrium concentration of point defects, the movement of dislocations is restricted almost entirely to the glide type. The glide plane is normally the plane with the highest density of atoms. For the wurtzite structure a glide often occurs in the (0001) basal plane along directions of the type $\langle 11\bar{2}0 \rangle$ [79, 81].

However, at higher temperatures an edge dislocation can climb out of its glide plane (in the wurtzite structure preferentially perpendicular to the basal plane), as schematically shown in Fig. 2.10(b). This process involves the creation, annihilation, and migration of point defects. Thus, it is thermally activated and occurs only at higher temperatures [79, 81].

2.6 Growth of free-standing GaN substrates – Hydride vapor phase epitaxy

Currently, the epitaxy of GaN-based devices had to be performed on extrinsic substrates, such as sapphire, silicon, or SiC. A high defect density arises in these devices because of the high lattice mismatch of GaN and the substrate materials. Although

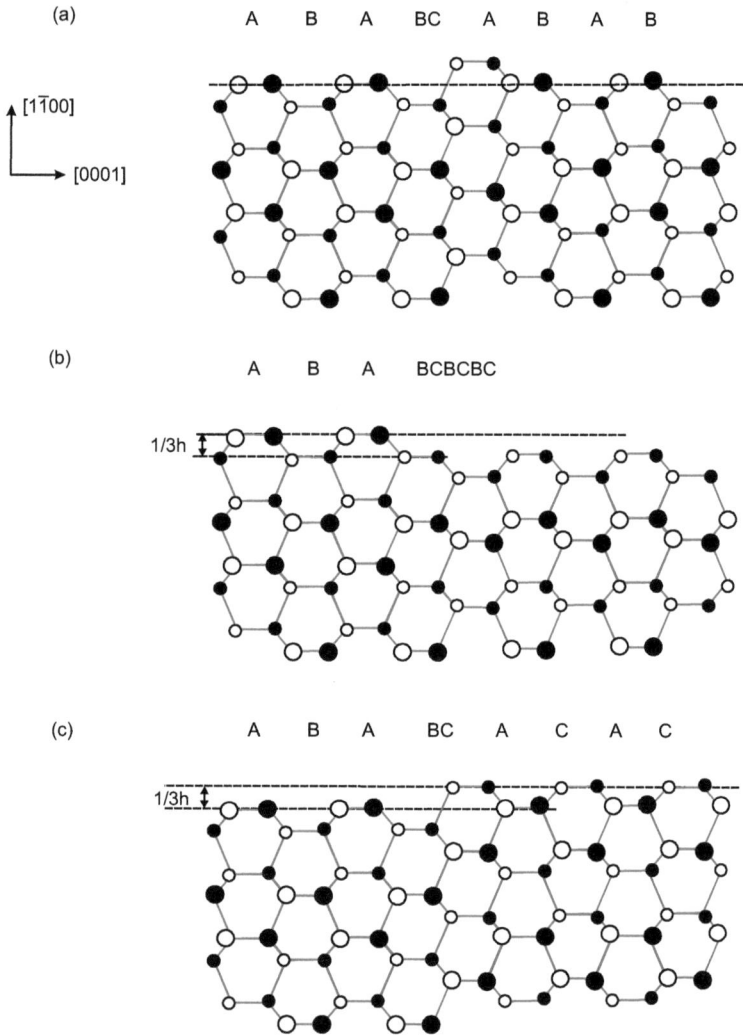

Figure 2.9: Schematics of stacking faults of type (a) $\{ABABCABAB\ldots\}$, (b) $\{ABABCBCBC\ldots\}$, and (c) $\{ABABCACAC\ldots\}$.

there are a lot of efforts made to reduce the defect density during the heteroepitaxy, only homoepitaxial growth can overcome this problem and improve the defect situation. Because of the very high partial pressure of nitrogen, GaN substrates cannot

2.6 Growth of free-standing GaN substrates – Hydride vapor phase epitaxy

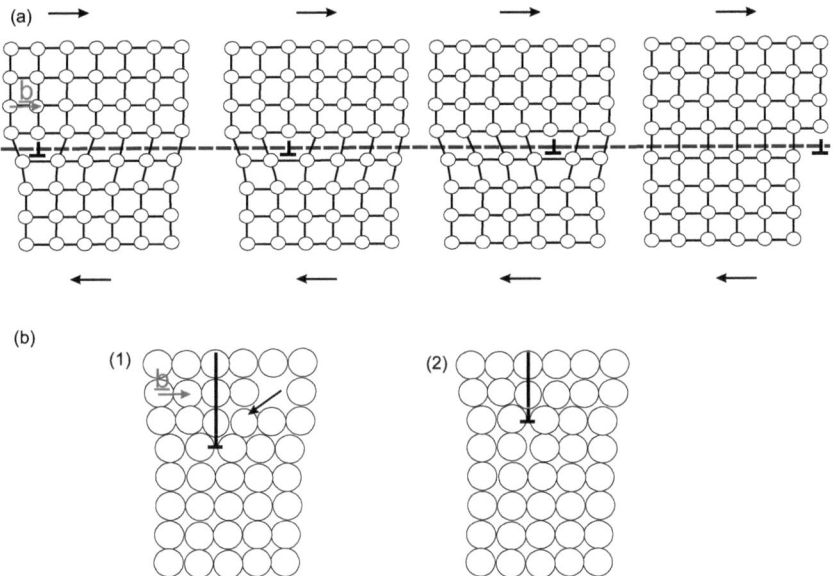

Figure 2.10: *Movement of an edge dislocation with the Burgers vector $\underline{b}$ [79]. The dislocation line is perpendicular to the diagram plane. (a) Dislocation glide: A dislocation moves to the right and eventually vanishes at the crystal surface. The arrows indicate the applied shear stress and the dashed blue line marks the glide plane. (b) Dislocation climb: (1) The vacancy in the lattice diffuse to the dislocation, producing a climb of the dislocation in (2).*

be extracted from the molten substance like in the Czochralski method, as used for common III-V semiconductors.

Using hydride vapor phase epitaxy (HVPE) free-standing (i.e. free from original sapphire substrate) GaN-substrates can be produced, where the concentrations of precursor related impurities and extended defects are reduced [78, 82]. In the HVPE reactor HCl gas is passed over a heated boat containing liquid Ga, forming GaCl vapor which is transported in a stream of hydrogen together with a separately injected flow of NH_3 over a heated sapphire substrate. The gases interact at the substrate surface and a film of GaN is deposited:

$$GaCl + NH_3 \rightarrow GaN + HCl + H_2$$

Typical growth temperatures are 1000°C and the growth rates vary from 1 to about 100 μmh^{-1} [83]. The stoichiometry can be controlled by adjusting the gallium boat temperature, the HCl flux, and the ammonia flow rate. Higher HCl gas flux leads to higher GaCl partial pressure and lower V/III ratio, and thus to a higher nucleation rate and a higher growth rate. The partial pressures of GaCl and NH_3 are e.g. 8×10^{-3} atm and 0.2 atm, respectively [84], and the typical V-III ratio varies between 25 and 1300 [78].

The process takes place in a quartz furnace tube and gallium is contained in an Al_2O_3 boat. After the growth of the sample the sapphire substrate can be removed by reactive ion etching or mechanical polishing, and a free-standing GaN substrate remains.

The dislocation density of such grown GaN substrates ranges from 1×10^6 cm^{-2} to 1×10^7 cm^{-2} [85, 86]. This is quite low, as compared with GaN films grown on Si or SiC substrates, but is still much higher than in the arsenide system. These high dislocation densities are associated with so-called threading dislocations. Threading dislocations are introduced in the initial stages of GaN/SiC growth, originating from misfit dislocations at the interface and propagate through the entire epitaxially grown layer, if they do not turn away from the growth direction.

Chapter 3

Scanning tunneling microscopy and spectroscopy

One of the most amazing inventions in physics of the late 20th century is the scanning tunneling microscope (STM). It was developed by G. Binnig, H. Rohrer, C. Gerber, and E. Weibel at IBM Zürich in 1981 [87, 88]. During an STM experiment a sharp conductive tip is scanned line-by-line over a conductive surface at a distance of less than a nanometer. Thereby a height map of a sample is created by the tip. This height map can be transformed into a color image, called STM image.

3.1 Theory of tunneling

In classical physics an electron cannot penetrate a potential barrier if its kinetic energy is smaller than the potential within the barrier. But in quantum mechanics this treatment of the barrier allows an exponentially decaying solution for the electron wave function in the barrier. The possibility of the electrons to pass this barrier is the so-called quantum mechanical tunneling effect.

3.1.1 One-dimensional tunneling

Tunneling from a tip into a sample through a vacuum barrier can be described by a simple one-dimensional model, as shown in Fig. 3.1. In the quantum mechanical description the one-dimensional time independent Schrödinger equation is used

$$-\frac{\hbar^2}{2m}\frac{d^2}{dz^2}\psi(z) + V(z)\psi(z) = E\psi(z). \tag{3.1}$$

The solution of this equation for the classical condition $E > V$ is the wave function

$$\psi(z) = \psi(0)e^{\pm ikz},$$

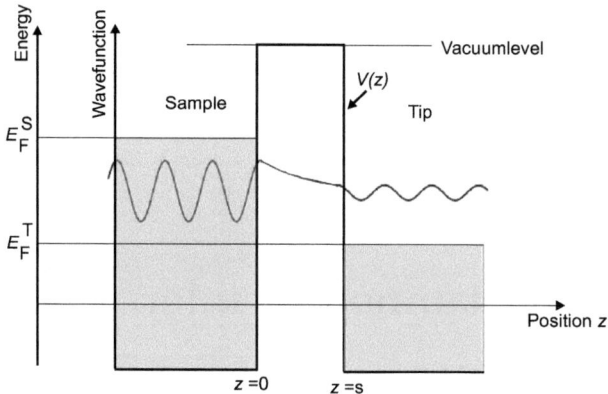

Figure 3.1: *The one-dimensional metal-vacuum-metal tunneling model.*

with the wave vector $k = \sqrt{2m(E-V)}/\hbar$ and z the distance perpendicular to the surface plane. In the classically forbidden zone the solution of the Schrödinger equation is

$$\psi(z) = \psi(0)e^{-\kappa z}, \qquad (3.2)$$

with the decay constant $\kappa = \sqrt{2m(V-E)}/\hbar$ and $\psi(0)$ the wave function at the left boundary of the barrier [89]. The probability density of observing an electron at the position z,

$$|\psi(z)|^2 = |\psi(0)|^2 e^{-2\kappa z}, \qquad (3.3)$$

decays exponentially.

This elementary model explains the tunneling of electrons through a vacuum barrier, as shown in Fig. 3.1. In this sketch a wave, coming from the left, is damped in the barrier and continues to propagate in the tip. Without any bias, however, there is no net tunneling current, due to the balancing of the opposite tunneling current contributions from the tip into the sample and from the sample into the tip. By applying a bias V, a measurable net tunneling current exists, which depends exponentially on the barrier width.

However, many factors are still unconsidered in this model, such as the geometry of the tip and the interaction between the tip and the sample. The Bardeen approach [90] presented in the following is a possible way to describe the interaction between the tip and the sample.

3.1.2 Bardeen approach

The Bardeen formalism uses the first-order time-dependent perturbation theory. An approach of the tip to the sample leads to an overlap of their wave functions, and thus a perturbation of the initial conditions. To perform perturbation calculations, a pair of subsystems with potentials $U^S(\underline{r})$ and $U^T(\underline{r})$ is defined, whereas S and T refer to the sample and tip, respectively. The sum of the two potentials of the individual system equals the potential of the combined system. The potentials of the subsystems disappear in the respective opposite system. These subsystems are separated by an interface drawn in Fig. 3.2 as a dotted line.

The unperturbed separated systems are described as

$$(-\frac{\hbar^2}{2m}\Delta + U^S)\Psi_\mu^S = E_\mu^S \Psi_\mu^S \tag{3.4}$$

$$(-\frac{\hbar^2}{2m}\Delta + U^T)\Psi_\nu^T = E_\nu^T \Psi_\nu^T. \tag{3.5}$$

Ψ_μ^S und Ψ_ν^T are the wave functions of the unperturbed states of the sample and the tip, respectively. The combined system is described through the Hamilton operator

$$H = H^S + U^T = -\frac{\hbar^2}{2m}\Delta + U^S + U^T. \tag{3.6}$$

When the subsystems start to overlap, perturbation theory can be applied for the combined subsystem by smoothly turning on the potential of the tip on the sample system. The time-dependent Schrödinger equation for this combined system state $\Psi_\mu(\underline{r},t)$ is

$$i\hbar\frac{\partial}{\partial t}\Psi_\mu = (-\frac{\hbar^2}{2m}\Delta + U^S + e^{\frac{\eta t}{\hbar}}U^T)\Psi_\mu. \tag{3.7}$$

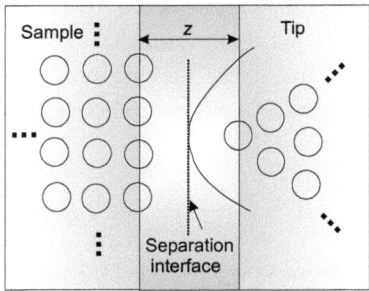

Figure 3.2: *Perturbation approach for quantum transmission.*

By summing over all the states in the tip ν and the sample μ, taking into account the occupation probabilities and the Fermi distribution, the tunneling current can be written as

$$I = \frac{4\pi e}{\hbar} \sum_{\nu\mu} [f(E_\mu^S - E_F^S) - f(E_\nu^T - E_F^T)] |M_{\mu\nu}|^2 \delta(E_\nu^T - E_\mu^S \pm eV), \quad (3.8)$$

with the applied bias V. The tunneling current is again the difference of the current from the sample into the tip and from the tip into the sample. The delta distribution arises from the elastic transition from a state μ in the sample to a state ν in the tip and reverse ($E_\mu = E_\nu$, energy conservation). The tunneling matrix element, which has a dimension of energy, describes the probability of the transition:

$$M_{\mu\nu}(\underline{r}, t) = \left\langle \Psi_\nu^T(\underline{r}, t) \left| U^T \right| \Psi_\mu^S(\underline{r}, t) \right\rangle. \quad (3.9)$$

However, the tunneling matrix element is still unconsidered. The description of the tip wave function is difficult, because the exact shape of the tip is mostly unknown. Simple assumptions of the tip shape and its wave function allow a determination of the tunneling matrix element. Such a model is the Tersoff-Hamann approximation [91,92], which is described next.

3.1.3 Tersoff-Hamann approximation

At the separation interface the potentials of the tip and the sample are equal to zero, and the time-independent Schrödinger equation is

$$[\Delta - \kappa^2]\Psi = 0 \quad \text{mit} \quad \kappa = \sqrt{\frac{2m}{\hbar^2} \frac{(\phi_S + \phi_T - eV)}{2}}, \quad (3.10)$$

with ϕ_S the work function of the sample and ϕ_T the work function of the tip. The tip is modeled geometrically and electronically isotropic and its contribution to the tunneling matrix element is given by an s-orbital wave function. The solution of the Schrödinger equation is an exponentially decaying wave of the topmost atom of the tip into the vacuum:

$$\Psi_\nu^T(\underline{r} - \underline{R}) \propto \frac{e^{-\kappa|\underline{r}-\underline{R}|}}{-\kappa|\underline{r}-\underline{R}|}, \quad |\underline{r} - \underline{R}| \neq 0, \quad (3.11)$$

with $\underline{R}$ the position of the topmost single atom of the tip and $\underline{r}$ a position in vacuum, and the tunneling matrix element then writes:

$$M_{\mu\nu} = -\frac{2\pi C \hbar^2}{\kappa m_e} \Psi_\mu^S(\underline{R}). \quad (3.12)$$

This Tersoff-Hamann tunneling matrix element only depends on the position of the topmost atom of the tip. The temperature dependence of the tunneling current is covered by the Fermi function:

$$f(E - E_\mathrm{F}) = \left[1 + \exp\left(\frac{E - E_\mathrm{F}}{k_\mathrm{B} T}\right)\right]^{-1}.$$

For $T = 0$ the Fermi function equals a step function. Using further the approximation $\frac{d\rho^\mathrm{T}}{dE} = 0$, where ρ is the density of states (DOS) (number of electrons per volume and energy at a given point in space and a given energy), the tunneling current is given by

$$I(\underline{R}, V) \propto \rho^\mathrm{T} \int_0^{eV} d\epsilon\, \rho^\mathrm{S}_\mathrm{lok}(\underline{R}, E^\mathrm{S}_\mathrm{F} + \epsilon), \tag{3.13}$$

whereas the local density of states (LDOS) ρ_lok is defined as:

$$\rho_\mathrm{loc}(\underline{R}, E) = \lim_{\epsilon \to 0} \frac{1}{\epsilon} \sum_{En = E - \epsilon}^{E} |\psi_n(\underline{R})|^2. \tag{3.14}$$

Thus, the tunneling current is given by the LDOS of the sample at the position of the tip. From equation (3.13) the differential conductivity is obtained by derivation to

$$\frac{dI}{dV} \propto \rho^\mathrm{T} \rho^\mathrm{S}_\mathrm{lok}(\underline{R}, E^\mathrm{S}_\mathrm{F} + eV). \tag{3.15}$$

The derivation of the tunneling current is directly proportional to the LDOS of the sample at the position of the tip and the energy $E_\mathrm{F} + eV$.

The influence of different tip wave functions on the tunneling current is further studied in Reference [93].

3.2 STM imaging and spectroscopy

3.2.1 STM operation modes

There are five parameters determinable or measurable in actual STM/STS experiments: x, y, z, I, and V. Using these parameters, the LDOS can be analyzed as a function of the position and the energy. Several STM-based microscopic and spectroscopic techniques [94] can be realized on semiconductor surfaces [95–101]. From the dependence of the tunneling current I on the distances z between the tip and the sample two different operation modes of the STM can be derived.

In the most common **constant-current mode** the tip is vertically adjusted in such a way that the current always stays constant by using a feedback loop ($z|_{I=I_\mathrm{set}}$).

During scanning a kind of a topographic image of the surface is generated by recording the vertical position of the tip $(z(x,y)|_{I=I_{\text{set}}})$.

The tunneling current does not only contain contributions from the Fermi level. In equation (3.13) the LDOS of the sample at the position of the tip has to be integrated over an energy range eV, implying that all included states contribute to the tunneling current. Constant-current images are thus images of the constant integral LDOS, weighted by the tunneling probability, which strongly depends on the energy: Electrons with lower binding energy values have higher tunneling probability, and electrons near the Fermi level have the highest tunneling probability.

In the **constant-height mode** the vertical position of the tip is held constant ($z = const$), equivalent to an interrupted feedback loop. The measured current as a function of lateral position $I(x, y)$ represents the surface image. Without a feedback loop STM images can be scanned very fast, up to several images per second. However, this mode is only appropriate for atomically flat surfaces as otherwise a tip crash would be inevitable.

The variation of the bias voltage allows the determination of the local electronic structure of the sample at the energy scale and the implementation of several spectroscopic modes.

One of the fundamental techniques used to investigate surface electronic structure is to acquire the so-called **point spectra** or I-V spectra. They contain information about the variation of the LDOS with the energy and, thus, about the surface states in semiconductors, band gaps, as well as doping. By interrupting the feedback loop ($x, y, z = const$) and varying the bias voltage V, the tunneling current $I(V)$ is measured. Additionally, the differential conductivity can be obtained as a function of bias voltage $\frac{dI}{dV}(V)$. This can, e.g., be performed by applying a small modulation on the tunneling bias and using a lock-in amplifier or by a numerical derivation.

Furthermore differential conductance maps of the surface, the so-called **STS images** ($\frac{dI}{dV}(x,y)|_{I=const}$) contain spatially resolved information about the LDOS of the sample at a dedicated energy. For this purpose the differential conductivity, measured using a lock-in amplifier, is determined additionally to a STM images, taken in the constant-current mode [102–104]. Scanning an STS image with measuring the differential conductivity gives truthful information about the LDOS of the sample at the dedicated energy as a function of the detected position. One difficulty in such experiments the adjustment of the frequency of the lock-in amplifier. The chosen frequency has to be well above the bandwidth of the feedback loop, so that the modulation of the tunneling current due to the modulation of the bias is not corrected by the feedback loop, resulting in a slow data acquisition.

One of the main problems associated with acquiring I-V spectroscopy data, particularly when studying semiconductors, is the sensitivity to very small currents. In the case of semiconductors, the current will approach zero as the bias approaches the edges of the band gap. This sensitivity limit can prevent the detection of features in the band gap and even the precise location of the band edges themselves. This limitation can be surmounted by taking advantage of the exponential dependence of the current on the width of the tunneling gap in the so-called **variable gap mode** [97], which is demonstrated in Fig. 3.3.

In general, the current increases by about an order of magnitude for every 1 Å decrease in the tip-sample separation. This means that if the feedback loop is opened and the tip moved towards the sample as the bias is lowered [Fig. 3.3(c)], the current will effectively be increased. After the data acquisition, however, the spectra have to be renormalized. The renormalization of tunneling spectra can be done by multiplication of the tunneling current by $\exp(2\kappa z)$, so that the effect of the additional movement of the tip toward the sample can be purged, provided that κ is known. Thus, the proper series of steps to perform when setting up spectroscopy data ac-

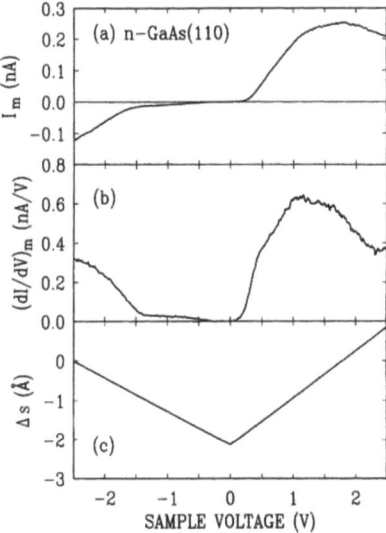

Figure 3.3: Variable gap mode. Measured data from an n-type GaAs(110) surface, showing (a) the tunnel current and (b) the differential conductance, as a function of sample voltage. The applied variation in tip-sample separation is shown in (c), from [97].

quisition is to acquire the so-called *I-z* spectra to measure the decay constant 2κ of the investigated material, then to obtain *I-V* data with a fixed tip-sample distance over the entire voltage range, and finally to acquire the variable gap data and compare it to the first set of *I-V* curves. A different method would be to calculated $(dI/dV)/(I/V)$. The last expression is independent on the tip-sample separation and is a good approximation for the LDOS of the sample [105].

The quantum mechanical nature of the tunneling phenomenon results in an exponential relationship between the tunneling current and the gap width $I \approx \exp(-2\kappa z)$. The value of κ, and hence the barrier height can be determined by acquiring ***I-z data*** $(I(z)|_{V,x,y=const})$. In this routine the feedback loop is opened and the tip is moved away or toward the sample while recording the tunneling current. These data will show an exponential decay constant equal to 2κ.

3.2.2 Influence of parallel momentum

Using $E = \frac{k^2}{2m} = \left(\frac{k_\perp^2 + k_\parallel^2}{2m}\right)$ the effects of nonzero parallel momentum $k_\parallel$ (see Fig. 3.4) can be included and the inverse decay length of the wave functions [106] can be rewritten as:

$$\kappa = \sqrt{\frac{2m}{\hbar^2}\frac{(\phi_S + \phi_T - eV)}{2} + k_\parallel^2}. \qquad (3.16)$$

Thus, the wave functions have the lowest decay constant when the parallel momentum is zero, corresponding to the $\overline{\Gamma}$ point in the surface Brillouin zone.

Since the tunneling probability decreases with parallel momentum, the vacuum tunneling barrier acts as a filter that passes mainly forward directed electrons. These

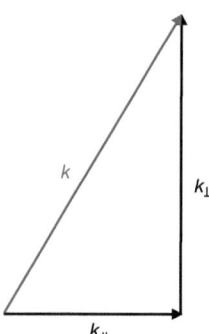

Figure 3.4: *Construction of k by vector addition of $k_\parallel$ and $k_\perp$.*

3.2 STM imaging and spectroscopy

electrons have relatively little parallel momentum and cannot couple with electronic states that lie far away from the surface Brillouin zone center $\overline{\Gamma}$. But, when the tunnel current increases, the barrier width decreases and the preference for forward directed electrons becomes weaker. Thus an increasing fraction of electrons has higher parallel momentum than the available states near the Brillouin zone center and thus increased tunneling from other points of the interface Brillouin zone occurs [107].

Tunnel currents from other areas in the Brillouin zone are detected in particular for semiconductors where either the high-lying occupied state disperses upward in energy as compared to the $\overline{\Gamma}$ point, or where the lowest energy unoccupied state disperses downward in energy as compared to the $\overline{\Gamma}$ point. Under these circumstances, energy conservation restricts tunneling at lower voltages to states with non-zero parallel momentum; at higher voltages, tunneling from all other states is possible, but nevertheless occurs preferentially from the $\overline{\Gamma}$ point.

For very sharp tips, this preferential tunneling at $\overline{\Gamma}$ is counteracted by momentum broadening through the uncertainty principle due to the strong lateral confinement of the tunneling electrons.

3.2.3 Tip-induced band bending

The tip-induced band bending (TIBB) is an additional effect which appears in STM experiments, when the vacuum energy of a semiconductor sample and the tip have different energetic positions [108–110]. The difference of the vacuum levels of the tip and the sample, caused by both a difference of the work functions and the applied

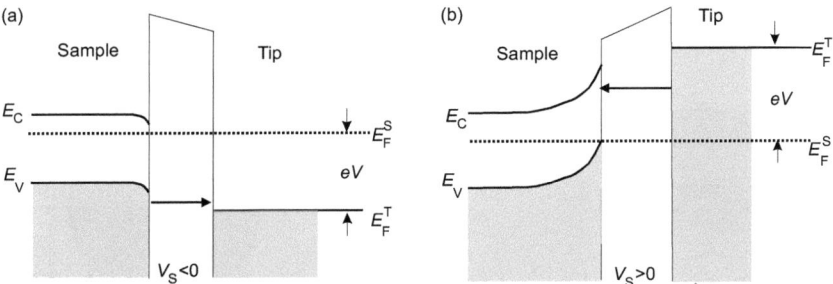

Figure 3.5: Energy-band diagrams for a metal insulator n-type-semiconductor system when (a) a negative sample bias and (b) a large positive sample bias is applied and an inversion layer is found.

bias, give rise to an electric field between the tip and the sample and thus to free carriers mainly on the surfaces of both materials [111].

For an n-type semiconductor, a negative sample bias with its electrical field between the tip and the sample leads to a local accumulation of electrons at the sample surface underneath the tip [Fig. 3.5(a)]. Since there are more electrons at the surface than in the bulk (apparent increasing of the n-doping), the potential energy of electrons at the surface is decreased and the bands bend downward [Fig. 3.5(a)]. Applying a small positive sample bias leads to a depletion of the electrons at the semiconductor surface underneath the tip (intrinsic conditions, not shown here). At larger positive biases a higher number of holes than electrons is present at the n-doped semiconductor surface, resulting in an inversion of the majority carrier and in a upward bending of the bands for a equilibrium carrier distribution [Fig. 3.5(b)].

3.3 Plan view and cross-sectional STM

The most important structural characterization method of semiconductor surfaces is STM, because of the possibility to take images with atomic resolution and thus to obtain directly the local structure.

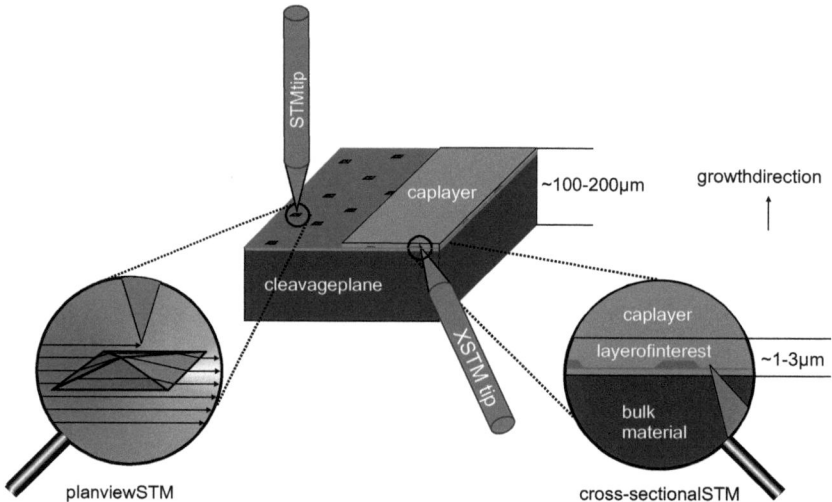

Figure 3.6: *Comparison of conventional plan-view STM with cross-sectional STM.*

Figure 3.6 shows the main difference between the conventional plan view STM and the cross-sectional STM (XSTM), which was mainly used in this work. In the plan view STM experiment e.g. a semiconductor surface or uncapped semiconductor heterostructures can be investigated on the growth surface, while in the XSTM experiment the sample is cleaved perpendicular to the epitaxial layers and thus buried semiconductor nanostructures can be studied. Such investigations of the capped nanostructures are of high interest, since during the capping of nanostructures different alloying and segregation processes may occur, determining the final spatial structure, which is used in optoelectronic devices.

In principle there is no big difference in performing both STM and XSTM experiments, except for the additional effort in XSTM for searching the layers of interest.

3.4 STM on III-V semiconductors

3.4.1 Contrast mechanisms

STM data contain both structural and electronic information of the investigated sample. Layers with a lower band gap usually appear brighter than those with a higher band gap. This electronic contrast occurs due to the different number of states electrons can tunnel into or out from for chemically different materials. In a similar way, nanostructures show different states, occurring from the quantum confinement. This modifies the tunneling probability and leads to an additional electronic contrast.

Surface steps and the atomic arrangement at the surface produce a direct structural contrast in the XSTM images, due to the exponential dependence of the tunneling current from the local distance between the tip and the sample. Strained QDs and QWs show an additional structural contrast in XSTM images occurring from their strain-induced relaxation after cleavage of the sample. Further information about the electronic and the structural contrast can be found in Refs. [71, 93].

3.4.2 Structural and electronic properties of the GaAs(110) surface

For the zincblende III-V semiconductors like GaAs, the {110} surface is the preferred cleavage plane, because it is the non-polar surface with the lowest density of dangling bonds. Non-polar surfaces consist of equal numbers of anions and cations and no electrostatic attraction exists between two parallel nearest-neighboring planes. The

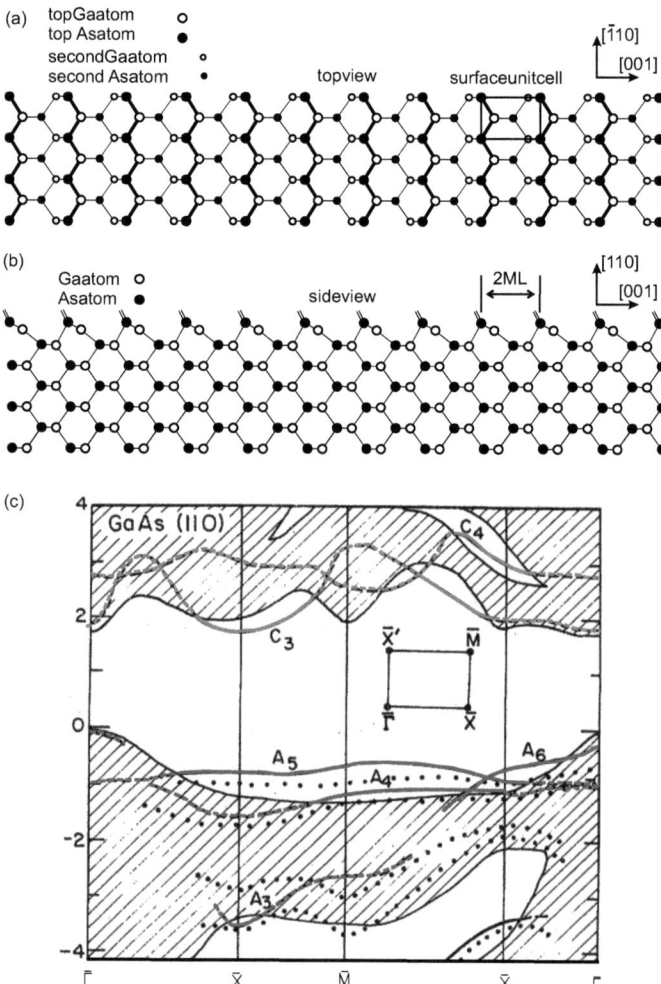

Figure 3.7: Realxed GaAs(110) surface. Schematic top view (a) and side view (b) of the relaxed GaAs(110) surface [71]. (c) Self-consistent pseodopotential calculation of the band structure of the relaxed GaAs (110) surface [112]. Projected bulk features are indicated by the shaded regions. Calculated surface bands are shown by solid lines in red for the anion derived states and in blue for the cation derived states. If the surface band becomes resonant, then a dashed line is drawn. The experimentally measured states are indicated by dotted lines.

3.4 STM on III-V semiconductors

charge neutrality of such surfaces does not require changes in stoichiometry or reconstruction.

The (110) surface is very similar to the (110) bulk plane. Although the lattice constant of the surface does not change as compared with the bulk lattice constant, there is still relaxation of the topmost atoms, as shown in Fig. 3.7(a) and (b). The surface layer consists of zigzag chains of alternating gallium and arsenic atoms. After cleavage each atom has one half filled dangling bond. After relaxation the electron in the dangling bond of the group-III atom is transferred to the dangling bond of the group-V atom, which is then completely occupied. This charge transfer is coupled with an outward motion of the anions, whereas the group-III atom moves towards the bulk.

The dangling bond states of the unreconstructed surface are energetically situated in the middle of the band gap. Due to the relaxation the empty surface states derived from the cations are pushed into the conduction band and the filled states, located at the anions, are shifted into the valence band [see Fig. 3.7(c)]. The surface features in the band diagram are labeled with A_i and C_i, depending on whether they are localized at the anion or at the cation, respectively.

In the XSTM experiment with positive sample bias electrons tunnel from the tip into the empty dangling bonds localized at the group-III atoms. Thus, the STM images the cationic sublattice and probes the conduction band in the spectroscopy mode. For negative biases, in contrast, the electrons tunnel from the occupied dangling bonds at the group-V atoms into the tip, imaging the anionic sublattice and probing the valence band.

3.4.3 Structural and electronic properties of non-polar GaN($1\bar{1}00$) and ($11\bar{2}0$) surfaces

For the wurtzite III-V semiconductors like GaN, the non-polar GaN$(1\bar{1}00)$ and $(11\bar{2}0)$ surfaces crystallographic planes are possible natural cleavage planes. Schematic models of these surfaces are shown in Fig. 3.8.

The two main relaxation mechanisms are a contraction of the GaN bond in the surface layer and a slight buckling and rehybridization of nitrogen and of gallium atoms. At the $(1\bar{1}00)$ surface the gallium and nitrogen atoms form an array of Ga-N dimers as indicated schematically in Fig. 3.8(a). The bond length of the Ga-N dimers on the $(1\bar{1}00)$ surface is 1.82 Å, corresponding to 6% contraction with respect to the bond length of the bulk (1.94 Å, see Tab. 2.1). The vertical displacement between nitrogen and gallium atoms in the surface dimer is 0.22 Å and corresponds

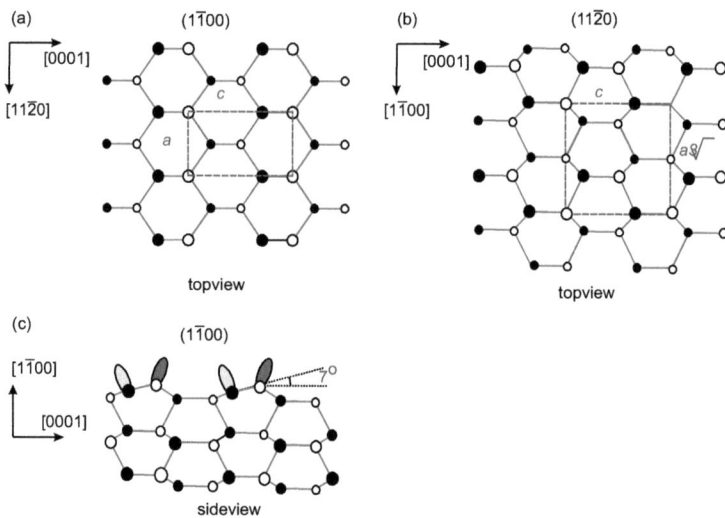

Figure 3.8: (a) Top-view of the GaN($1\bar{1}00$) surface, (b) top-view of the GaN($11\bar{2}0$) surface. The black circles outline gallium atoms, while the white circles represent nitrogen atoms. The dashed rectangles show the surface unit cells. (c) Side view of the GaN($1\bar{1}00$) surface relaxation. Electrons of dangling bonds transfers from gallium atoms to nitrogen atoms. This causes the relaxation of the surface. The buckling angle is 7°, from [113].

to a buckling angle of 7° [see Fig. 3.8(c)] [114]. The bond angles around the gallium atom become 118°, 118°, and 113°, while those around the nitrogen atom are 105°, 105°, and 114° [114].

The structure of the $\left(11\bar{2}0\right)$ surface contains meandering chains of threefold-coordinated gallium and nitrogen atoms, as indicated in Fig. 3.8(b). In each unit cell there are four surface atoms, two gallium and two nitrogen atoms. The calculated Ga-N bond lengths in the surface chain are 1.85 Å, corresponding to contractions of 4% compared to the bulk value. The vertical separation between the nitrogen and gallium atoms in the surface layer is 0.22 Å, the same value as found for the $\left(1\bar{1}00\right)$ surface. In this case, the nitrogen atoms exhibit bond angles of 107°, 106°, and 101°. The gallium atoms again relax towards an sp^2 configuration with bond angles of 119°, 116°, and 115° [114].

The surface energy for the $\left(1\bar{1}00\right)$ surface is found to be 1.95 eV (two-atom cell), corresponding to 118 meV/Å^2. For the $\left(11\bar{2}0\right)$ surface the surface energy is 3.50 eV/(four-atom cell) which corresponds to 123 meV/Å^2 [114]. It is important to note that although the energy per surface atom is slightly higher for the $(1\bar{1}00)$ surface, the energy density of the surface is slightly higher on the $(11\bar{2}0)$ surface because of the 15% higher surface atom density. The calculated surface energies for GaN may be compared with the one of GaAs(110), which is 1.20 eV/(two-atom cell) corresponding to 54 meV/Å² [114]. The larger cleavage energy for GaN results from two effects: the density of bonds that are broken to create the surface is larger in the case of GaN, and the energy required to break each bond is higher (see Tab. 2.1).

The calculation of the electronic structure for the $(1\bar{1}00)$ surface was performed by several groups using the density functional theory (DFT) in the local-density approximation (LDA) [114] and the pseudopotential-DFT method [115]. In such LDA calculations for bulk GaN a band gap of about 1.9 eV compared with the experimental gap of 3.4 eV was obtained. This deviation is typical for LDA calculations of semiconductors [116].

The calculated LDA electronic structure for both the ideal and the relaxed $(1\bar{1}00)$ surface is shown in Fig. 3.9(a). The occupied surface state (S_N) is located at the nitrogen derived dangling bond while the unoccupied surface state (S_{Ga}) is a dangling bond localized at the gallium atom. For the fully relaxed structure, the nitrogen-derived band lies just below the valence-band maximum, while the gallium-derived band lies just above the conduction-band minimum, thus the LDA predicts no intrinsic surface states in the gap for the $(1\bar{1}00)$ surface.

PDFT calculations of the electronic structure and the corresponding density of states for the $(1\bar{1}00)$ surface are shown in Fig. 3.9 (b) and (c), respectively. Unoccupied gallium dangling bond states are found at 0.7 eV below the CBM for the $(1\bar{1}00)$ surface. These surface states were made responsible for the Fermi-level pinning on n-type GaN. In contrast, the occupied surface states behave quite differently: These states are associated with dangling bonds on the nitrogen atom and overlap with the valence band, i.e., they do not create levels within the band gap.

For the $(11\bar{2}0)$ surface PDFT calculations yield similar results as for the $(1\bar{1}00)$ surface. Two unoccupied gallium dangling bond states are calculated at 0.5 eV below the CBM, whereas two occupied surface states are calculated to be located within the valence band [115].

3.4.4 STM images of non-polar wurtzite surfaces

Before this work no atomically resolved STM results were published for cleaved wurtzite GaN surfaces. Only for the analog non-polar wurtzite surfaces CdSe$(10\bar{1}0)$, CdSe$(11\bar{2}0)$ and CdS$(10\bar{1}0)$ STM images were published [117]. Figure 3.10 shows

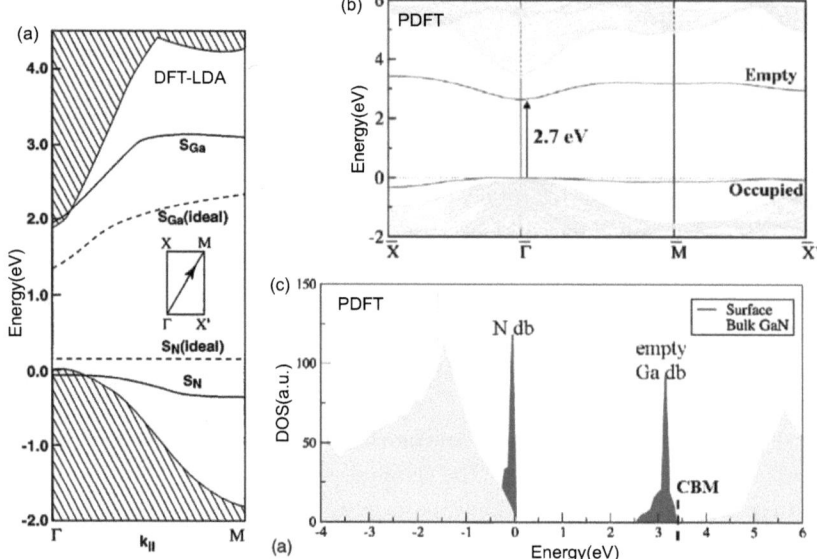

Figure 3.9: GaN($1\bar{1}00$) surface band structure. (a) Electronic band structure calculated within the density functional theory in the local-density approximation for both the ideal (dashed lines) and relaxed (solid lines) GaN($1\bar{1}00$) surfaces of GaN. The shaded region corresponds to the bulk projected band structure. The dashed lines indicate the dispersion of the surface states calculated for the surface without structural relaxation [114]. Pseudopotential density functional theory calculations of (b) electronic band structure and (c) density of states of the relaxed surface. Gray lines indicate the projected bulk band structure. The zero of energy is set at the bulk valence band maximum. Relevant energy differences between the valence band maximum and surface states (in red) are indicated by an arrow [115].

atomically resolved STM images of the occupied [(a1) and (b1)] and empty [(a2) and (b2)] states of the two non-polar CdSe cleavage surfaces. For comparison the respective images obtained on the GaAs(110) surfaces are shown in Fig. 3.10(c1) and Fig. 3.10(c2).

The STM images of the $(11\bar{2}0)$ surface in Fig. 3.10(a1) and Fig. 3.10(a2) show zigzag chains along the [0001] direction in both the occupied and empty state images. These zigzag chains are separated from each other in the $[10\bar{1}0]$ direction by 0.74 nm, and have a periodicity in the [0001] direction of 0.78 nm. The corresponding surface unit cell of the CdSe$(11\bar{2}0)$ surface is indicated in Fig. 3.10(a1). The width of the zigzag chain (measured perpendicular to the chain direction from the maxima on the

3.4 STM on III-V semiconductors

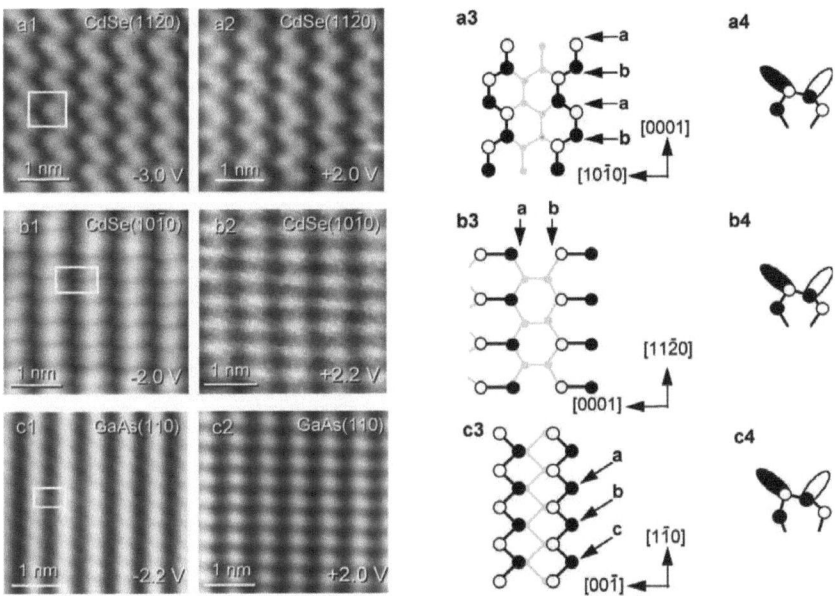

Figure 3.10: Occupied (frames labeled 1, column on the extreme left) and empty (labeled 2, second column from left) state images as well as a schematic view (3, third column) and side view (4, fourth column) of the wurtzite (11$\bar{2}$0) [frames labeled (a), first row] and (10$\bar{1}$0) (b) cleavage surface of CdSe (second row) and the cubic GaAs(110) surface [frames (c), third row]. The filled and empty ellipses in the fourth column represent a filled and an empty dangling bond, respectively, from [117].

left-hand side of the chain to those on the right-hand side of the chain) is 0.23 nm. The surface unit cell, obtained from the STM images shows a (1×1) relaxation. These experimental findings are in good agreement with the theoretical considerations, shown in Fig. 3.10(a3) and Fig. 3.10(a4).

The STM images of the CdSe$(10\bar{1}0)$ surface exhibit a rectangular pattern and again a (1×1) unit cell. The dimensions of the surface unit cell shown in Fig. 3.10(b1) are (0.43×0.70) nm². The longer side of the surface unit cell is oriented parallel to the [0001] direction. These experimentally observed structural results are in good agreement with the theoretically predicted calculations [118], shown in Fig. 3.10(b3) and Fig. 3.10(b4). No tunneling was possible for biases between about -1.8 and $+1$ V, indicating that no surface states exist in the band gap (2.8 eV wide).

In summary, atomically resolved STM investigations of non-polar CdSe cleavage surfaces confirm the (1 × 1) relaxation for both the $(11\bar{2}0)$ and $(10\bar{1}0)$ cleavage surfaces. The image features can be attributed in the occupied and empty state images to the completely filled and empty dangling bonds above the anions and cations, respectively. This is the signature of a charge transfer from anions to cations in analogy to the GaAs(110) surface [Fig. 3.10(c1)-(c4)].

Chapter 4

Experimental setup

4.1 The UHV chamber system

STM measurements in this work were performed using a non-commercial microscope. If the STM experiment would be done with air exposure, the sample surface would oxidize immediately (within nanoseconds), making pristine investigations of the sample surface impossible. Thus, the STM unit needs to be located in ultra high vacuum (UHV).

The used UHV system (see Fig. 4.1) consists of a preparation chamber, a fast entry, and an STM chamber. The fast entry chamber (load lock), which is connected to the preparation chamber, can be used to introduce tips and samples into the STM chamber without destroying the UHV. The chambers are separated by valves. The tips and samples can be moved between the chambers via magnetic transfers. Within the preparation chamber a storage for 32 tips and the tip heater are located, allowing an electron bombardment of the tips for preparation. Furthermore, the magnetic transfer in the preparation chamber allows storaging of five samples. Within the STM chamber six backup tips can be stored and exchanged for the current measurement without opening the valve to the preparation chamber. For the manipulation and transportation of tips and samples two wobble sticks, located in the STM and the preparation chamber, are used.

For establishing of UHV several pumps are installed: a diaphragm and a turbomolecular pump connected to the load lock, a titanium and an ion getter pump in the STM chamber, and an erbium sublimation and an ion getter pump in the preparation chamber. Moreover for elimination of water molecules the whole system has to be baked at a temperature of 120°C for several days. The experiments in this work

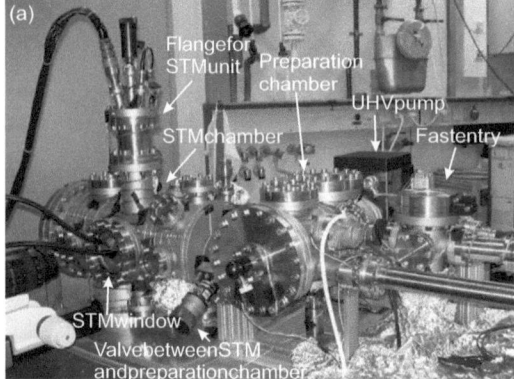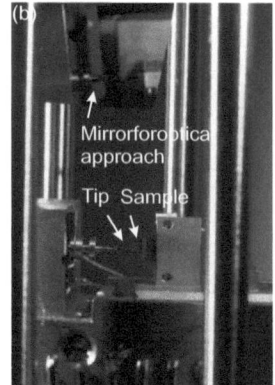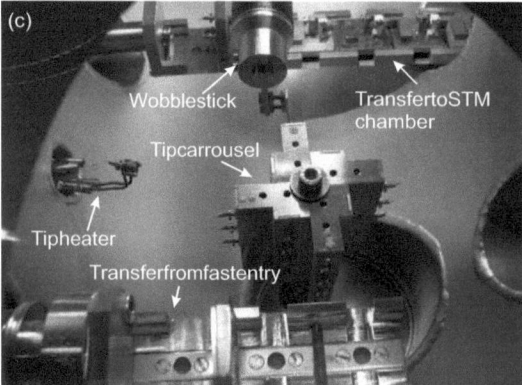

Figure 4.1: *Photographs of XSTM setup. (a) UHV chamber system (STM chamber, preparation chambers and fast entry). (b) Side view on the tip-sample region. (c) Look inside the preparation chamber. (d) Sample holder used for XSTM investigations.*

were performed at a base pressure below 2×10^{-8} Pa, enabling STM measurements of cleaved sample surfaces without significant contamination for about one week.

Additional details to the construction and the setup of the UHV system can be found in references [71, 72, 119].

4.2 Tip preparation

The sharpness and chemical stability of the tips used in STM are very important factors, affecting the resolution of the observed images. STM tips are typically made from metal wires of tungsten or platinum-iridium with a diameter around 0.2 mm. In this work commercial electrochemically etched platinum-iridium as well as tungsten tips were used, whereas the latter are prepared as described in the following. Using a 10% NaOH solution the tips are etched by applying a voltage between the tungsten wire as anode and a stainless steel cathode. The etching process involves the oxidative dissolution of tungsten to soluble tungstate anions WO_4^{2-} at the anode with a maximum etching velocity directly underneath the meniscus [120].

In order to maintain a good vacuum the tips have to be cleaned by several cleaning steps in deionized water and ethanol to remove the oxide layer covering the tip. In the preparation chamber the tips can be finally cleaned by electron bombardment and transferred into the STM chamber to perform measurements.

4.3 Sample preparation

The samples investigated in this work are grown on substrates with a typical thickness between 300 and 500 μm. According to experience, the cleavage of such rather thick samples causes a high step density of the corresponding cleaved surface. In order to minimize this surface step density the samples are grinded down mechanically to about 200 μm for GaAs and to about 100 μm for GaN substrates.

The thinned pieces of about 4×5 mm size are provided with a notch of about 2 mm length or Knoop indenter, indicating the possible later cleavage plane. Afterwards the sample pieces are glued on copper plates and an indium contact between the sample and the copper plate is made. The glued samples are build into sample holders and positioned in the load lock for transferring into the UHV system. A photograph of a sample in an XSTM sample holder is shown in Fig. 4.1(d).

4.4 Cleavage of the sample and tip approach

The cleavage of the sample occurs in the STM chamber under UHV conditions. For this purpose the sample holder has to be moved behind the STM unit, so that the sample area including the notch can be pushed against the STM unit and the sample can be cleaved.

Afterwards the cleaved sample and a cleaned tip can be positioned in the STM unit and the sample can be approached by an inertial walker to the tip, controlled through an optical microscope. The sample is moved to the tip until its reflection can be seen on the sample surface. Afterwards an automatic approach is started, consisting of repeated walker steps and tip extensions. Thereby after every single walker step the tip will be extended to check if a contact between the tip and the sample is obtained. After establishing of the tunneling contact the measurement can be started and the grown layers investigated.

Chapter 5

Structural and electronic properties of diluted GaAsN

In recent years diluted III-V nitride semiconductors attracted wide scientific attention because of both their particular electronic properties as compared with usual arsenide compounds and potential device applications. Their uniqueness is due to a large band bowing of the band gap by addition of minute amounts of nitrogen up to 5% to GaAs, inducing broad perspectives for band-gap engineering [42]. This band gap shift occurs due to a strong interaction between the conduction band of GaAs and a narrow resonant band formed by the nitrogen states, resulting in a splitting of the conduction band of GaAsN and thus a rather strong reduction of its fundamental band gap, as compared to GaAs [46], as discussed in detail in section 2.2.

However, the optical emission of GaAsN/GaAs QWs typically shows a strong degradation with increasing nitrogen content [121, 122]. This intensity decrease is often attributed to an incorporation of crystal defects during heteroepitaxial growth and focuses therefore the interest on a structural characterization on the atomic scale. Furthermore, the device performance has not reached expectations because of the reduction of the carrier mobility and thus the reduction of the device efficiency [62,123]. Thus, the technological importance of these alloys is still margined. In order to understand the processes behind both crystal defects and carrier mobility limitation, intensive theoretical work has been performed, even though their experimental verification is still lacking.

In this context, GaAsN/GaAs quantum wells were investigated here by XSTM and XSTS to study both the local atomic arrangement and the related electronic structure.

5.1 Sample structure and growth

In this work two different samples containing GaAsN layers with different nitrogen concentration were used. Both samples were grown by O. Schumann [124] on a GaAs(001) substrate in a VG V80H solid source MBE chamber, assisted by an Oxford applied research radio-frequency nitrogen plasma source. For the nitrogen incorporation the nitrogen plasma was ignited far below the GaAsN layer in order to stabilize the plasma cell at equilibrium conditions. During the growth of the GaAsN layers the shutter in front of the burning nitrogen plasma source was open, while it was kept closed as the surrounding GaAs was grown. The nitrogen concentration for the GaAsN layer was varied with the N_2 flow rate. In order to determine the effectively incorporated nitrogen concentration several samples with different plasma source parameters were grown. Their nitrogen contents were determined using X-ray diffraction measurements [124, 125].

The first sample studied here contains a nominally 15 nm thick GaAsN layer with a nominal nitrogen concentration of 1.2% [Fig. 5.1(a)]. The sample growth was

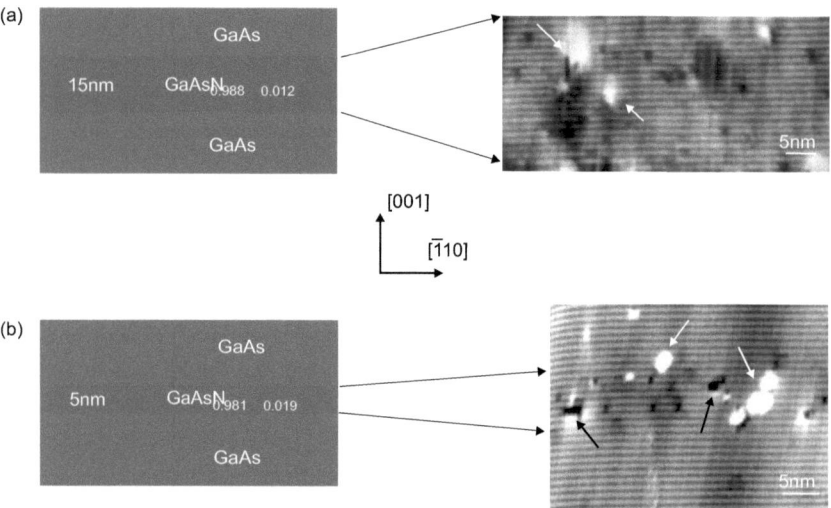

Figure 5.1: The sample structure (left) and the respective XSTM image (right) of (a) a nominally 15 nm thick $GaAs_{0.988}N_{0.012}$ layer, acquired at -2.5 V and 70 pA, and (b) a nominally 5 nm thick $GaAs_{0.981}N_{0.019}$ layer, acquired at -3 V and 90 pA. Adsorbates are indicated by white arrows, whereas cleavage-induced crystal imperfections are indicated by black arrows.

performed at a constant substrate temperature of 490°C. For the second GaAsN/GaAs sample a nominal nitrogen concentration of 1.9% was incorporated and the sample growth was performed at 400°C. In order to reduce strain-related imperfections of the GaAsN layer, the thickness of this layer was reduced to 5 nm [Fig. 5.1(b)]. Further details of the material growth are reported in Refs. [124, 125].

5.2 Nitrogen incorporation into GaAs

In Fig. 5.1(a) and (b) on the right side filled states XSTM images are shown of the $GaAs_{0.988}N_{0.012}$ and $GaAs_{0.981}N_{0.019}$ layers, respectively. The lines running parallel to the $[\bar{1}10]$ direction originate from the arsenic rows on the GaAs(110) surface. In these XSTM images several dark spots within the arsenic rows can be identified [126–128]. These features were counted directly from the XSTM images and their average concentration was determined to $(1.20 \pm 0.05)\%$ and $(1.90 \pm 0.08)\%$ for the first and the second sample, respectively. Thus, this measured concentration is in excellent agreement with the nominal nitrogen concentration and thus single nitrogen atoms can be identified in the XSTM images within the topmost layer of the cleaved GaAs(110) surface [126–128].

Also a few adsorbates, indicated in Fig. 5.1(a) and (b) by white arrows, and cleavage-induced defects, indicated in Fig. 5.1(b) by black arrows, are observed. As compared with the surrounding GaAs the density of adsorbates is increased for the GaAsN layers. Moreover, the contrast in the GaAsN layers is darker than in the surrounding GaAs. This dark contrast originates from the strain-induced inward relaxation of the tensilely strained GaAsN layers, further supporting the detection of the layers.

5.3 Identification of single nitrogen atoms

In the following the identification of single nitrogen atoms will be discussed using XSTM results of the sample containing 1.2% nitrogen. Figure 5.2 shows several high resolution filled state XSTM images of nitrogen-related features. In all images the anion sublattice is imaged. Some of the anions appear to be darker than others. These darker features were associated with nitrogen atoms, which are substitutional for the arsenic sublattice [126–129]. This appearance of dark spots corresponding to single nitrogen atoms occurs, however, in two gray levels, as shown in Fig. 5.2(a). The darkest spots marked by black circles correspond to nitrogen atoms directly in

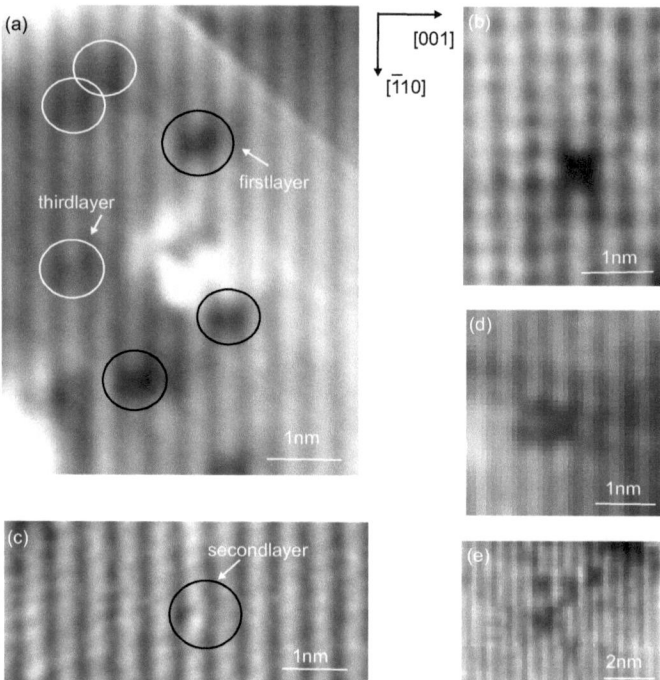

Figure 5.2: Detailed filled state XSTM images (taken at sample biases between –2.5 V and –3 V) of (a) first- and third-layer nitrogen related features, (b,c) two single depressions within the arsenic chain, corresponding to (b) a single nitrogen atom in the first plane and (c) a second-layer nitrogen atom, (d) two single depressions within two neighboring arsenic chains, corresponding to a pair of nitrogen atoms, oriented in the [001] direction; (e) several dark spots within arsenic chains, corresponding to a group of nitrogen atoms.

the surface layer (first layer), as further schematically shown in Fig. 5.3. Similar features with a reduced contrast indicate substitutional nitrogen in the third atomic layer below the surface, as schematically shown in Fig. 5.3. These third-layer nitrogen atoms are marked in Fig. 5.2(a) by white circles. An XSTM image of a single nitrogen atom incorporated in the first layer is shown in Fig. 5.2(b). This image as well as the characterization of an incorporated nitrogen atom will be discussed in detail in the next section.

Second-layer nitrogen atoms are characterized by a kink of the arsenic chain, where the arsenic atom is slightly shifted along the [001] direction [126, 127], as marked in Fig. 5.2(c) by a circle and schematically shown in Fig. 5.3.

5.4 Nitrogen versus arsenic vacancy in the GaAs(110) surface

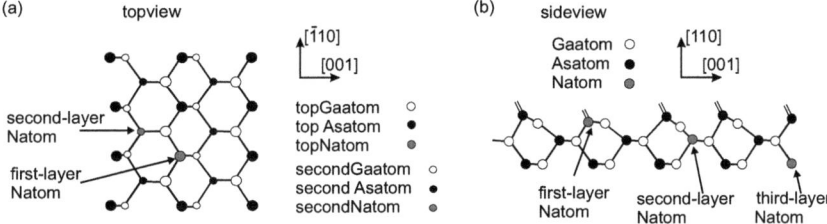

Figure 5.3: (a) Schematic top view of the GaAsN(110) surface showing first-layer atoms (large dots) and deeper-layer atoms (small dots). A smaller nitrogen atom (green dot) is substituted for a first- and a second- layer arsenic atom. (b) Schematic side view of the GaAsN(110) surface. A first-, second-, and third-layer nitrogen atom is incorporated.

The nitrogen distributions in the different layers do not show any obvious ordering. The nitrogen arrangement rather appears to have a random distribution for such low concentrations. Nevertheless, locally pairs or even groups of nitrogen are occasionally observed, as shown in Fig. 5.2(d) and (e), respectively.

5.4 Nitrogen versus arsenic vacancy in the GaAs(110) surface

In this section the appearance of the dark features will be studied in detail in order to distinguish nitrogen atoms from arsenic vacancies or other defects producing a similar contrast in XSTM images. It should be noted that the concentration of the observed arsenic vacancies is found to be $(0.120 \pm 0.017)\%$, one order of magnitude smaller than the nitrogen concentration.

Fig. 5.4(a) shows a filled state XSTM image of a single nitrogen-related feature. As visible in this image, there is a localized reduction in the tip height directly in the arsenic row, suggesting that a single arsenic atom has been replaced by a deeper-lying nitrogen atom. The structure of a nitrogen atom can better be seen in the height profiles through the same filled-state images, as shown in Fig. 5.4(b). This height profile (K_1-K_2) along the [001] direction demonstrates, that at the position of the nitrogen atom, corresponding to the dark region in the XSTM image, the tunneling probability is strongly decreased as compared to the one at the positions of arsenic atoms, while a small bump is still visible. These structural findings are in good agreement with simulated STM images, based on *ab-initio* calculations of the GaAsN surface, as shown in Fig. 5.4(c). This calculated STM image at -3 V shows the filled

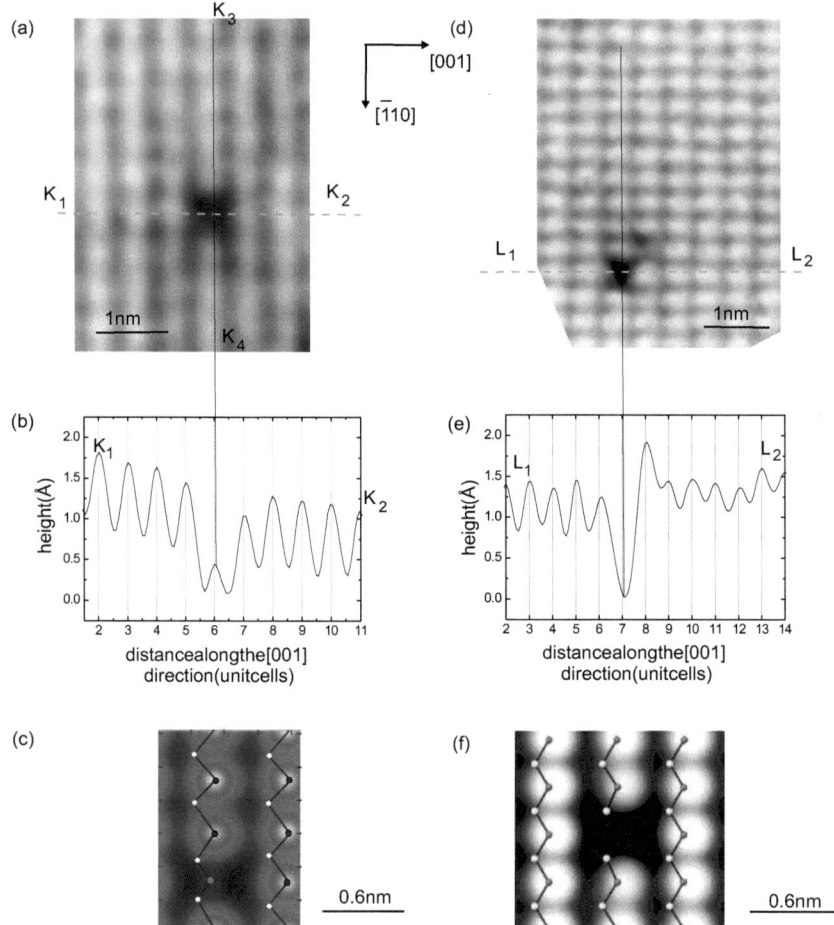

Figure 5.4: (a)-(c) A nitrogen atom within the GaAs(110) surface: (a) Filled state XSTM image of a single dark depression within the arsenic chain, (b) height profile taken along the line K_1-K_2, and (c) ab-initio pseudopotential calculated filled states STM image of the GaAsN surface at −3 V [130]. (d)-(f) Single arsenic vacancy within the GaAs(110) surface: (d) Filled state XSTM image of a single depression within the arsenic chain, (e) height profile taken along the line L_1-L_2, and (f) ab-initio pseudopotential calculation of a filled state STM image at −1.0 V [131].

states of the GaAsN surface with a clear local tip-height decrease at the position of the nitrogen atom. Furthermore, arsenic atoms in the same $[\bar{1}10]$ chain adjacent to this nitrogen-related feature appear to be slightly and symmetrically depressed along the K_3-K_4 line direction in Fig. 5.4(a).

Nitrogen atoms appear very similar to arsenic vacancies within the GaAs(110) surface, as shown in Fig. 5.4(d), since nitrogen is much smaller than the arsenic atom it replaces in the lattice. The arsenic vacancy also appears as a single depression in the filled-state image. The structural deformation is well localized around the vacancy position. These structural findings are again in good agreement with theoretical *ab-initio* calculations, as shown in Fig. 5.4(f). A closer look at the height profiles along the [001] direction [Fig. 5.4(b) and (e)] taken above both the nitrogen atom and the single vacancy provides their distinctiveness. An additional height minimum in the occupied DOS of the GaAs(110) surface can be clearly recognized for both. But the arsenic vacancy is located much deeper in the surface, as compared with the nitrogen atom. Moreover, at the position of the nitrogen atom the tip height is locally increased, confirming that a smaller atom is located within the arsenic chain. Finally it should be noted that a nitrogen correlated feature can be distinguished from other crystal defects such as impurities or interstitials, since they appear differently in XSTM images [132].

5.5 Local density of states

5.5.1 STS results

In the GaAsN layer with a nitrogen concentration of 1.2%, shown in Fig. 5.1(a), current-voltage spectra as well as conductivity-voltage spectra were taken. The latter spectra were measured with a lock-in amplifier, using the variable gap mode [97] with a Δz change of 1.5 Å per 1 V, an oszillator frequency of 10 kHz, an oszillator amplitude of 40 mV, and a time constant of the low-pass filter of 10 ms.

Figure 5.5(a) shows a diagram of the logarithmically displayed tunneling current I and Fig. 5.5(b) the normalized differential conductivity $(\mathrm{d}I/\mathrm{d}V)/\overline{(I/V)}$ as a function of the sample voltage. The $(\mathrm{d}I/\mathrm{d}V)/\overline{(I/V)}$ curve is calculated after reference [133], using $\overline{(I/V)} = \sqrt{(I/V)^2 + c^2}$ with $c = 5 \times 10^{-11}$ A/V. Both curves demonstrate clearly the energetic positions of the conduction band minimum E_C and the valence band maximum E_V. The STS measurements were performed at a GaAsN layer, containing a large number of adsorbate atoms with a concentration of up to 10^{13} cm^{-2}, leading to a strong Fermi level pinning and hindering the band bending. The observation of the Fermi energy E_F close to the center of the band gap can be related to the energetic

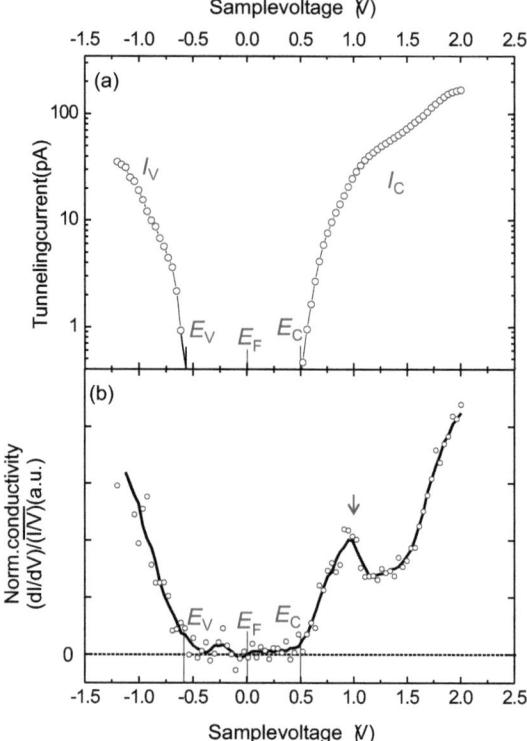

Figure 5.5: (a) Diagram of the logarithmically displayed tunnel current and (b) normalized differential conductivity $(dI/dV)/\overline{(I/V)}$ as a function of the sample voltage.

position of these defects and/or to the position of the investigated sample area in between n- and p-doped regions.

At negative voltages the tunneling current as well as the differential conductivity illustrate the valence band states, if the valence band edge of the sample is energetically above the Fermi level of the tip. At positive voltages the tunneling current as well as the differential conductivity arise from the conduction band states, if the Fermi level of the tip is energetically above the conduction band edge of the surface. At about -0.6 eV and 0.5 eV the onset of the valence and the conduction band DOS are visible, respectively. Thus, the band gap at this surface can be estimated to be about (1.1 ± 0.2) eV, considerably differing from the fundamental band gap of GaAs (1.43 eV) by about 0.3 eV.

Moreover, in comparison with the characteristic DOS of GaAs, the normalized differential conductivity of GaAsN demonstrates an additional feature, indicated in Fig. 5.5(b) by a red arrow. There is a peak at about 0.5 eV above of the conduction band edge which can be attributed to the nitrogen. The appearance of nitrogen states above the conduction band edge has also been observed previously by STS, being attributed to its acceptor-like level [134].

In order to understand the origin and the characteristics of this nitrogen-induced feature in detail, the measured DOS is now compared with theoretical calculations.

5.5.2 Comparison with the BAC model calculations

The simplest model used to describe the very strong band gap bowing in diluted GaAsN is the band anti-crossing (BAC) model [46], discussed in section 2.2 and shown in Fig. 2.2(a) for the case of $GaAs_{0.988}N_{0.012}$.

In Fig. 5.6 the dispersion from the BAC model is compared with the measured DOS. First of all, the reduction of the fundamental band gap of GaAs, from 1.43 eV to about 1.1 eV is clearly visible for both curves. Moreover, both the BAC model energy dispersion and the measured DOS exhibit three different contributions. For negative sample biases the tunneling current originates from the valence band states, showing a good agreement with the energetic position of the valence band in the BAC energy dispersion despite of a slight shift to higher energy. At positive sample biases two different contributions to the tunneling current can be identified. First, the onset of the conduction band can be found at about 0.5 eV, corresponding to the states of the lower E_- band at the wave vector $k = 0$. Further, the measured curve exhibits a peak at about 1 eV, which can be related to the states of the lower E_- band at higher wave vectors. This flat part of the E_- band dispersion corresponds to a high density of states, resulting in the measured peak. At sample voltages higher than 1.25 eV the tunneling current mainly results from the states of the upper E_+ band.

It can clearly be seen, that the rather simple BAC model nicely provides the position of the band edges of GaAsN at a given nitrogen concentration. But it fails to provide the correct DOS. In the BAC model the perturbed states have a well-defined wave vector k, leading to an infinite number of states and thus an infinite DOS. In order to solve this problem, J. Wu et al. [135] and M.P. Vaughan et al. [136] used Green's functions to describe the conduction band states after mixing.

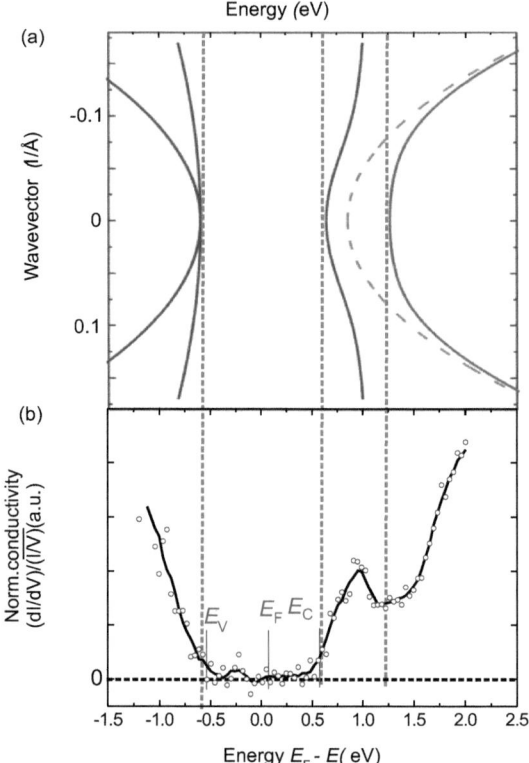

Figure 5.6: (a) Valence-band and conduction-band dispersion relations for GaAs$_{0.988}$N$_{0.012}$ at 300 K calculated after the BAC model, compared with (b) the measured normalized differential conductivity $(dI/dV)/\overline{(I/V)}$ from Fig. 5.5(b).

5.5.3 Green's function approach

A key point of this model is that the nitrogen energy level obtains an imaginary component. This can be associated with a finite lifetime on the nitrogen state and gives rise to homogeneous broadening of the energy level, as schematically shown in Fig. 5.7.

The density of states at energy E in the dilute nitride can be found from the imaginary part of the Green's function $G(E, E_+)$ via

5.5 Local density of states

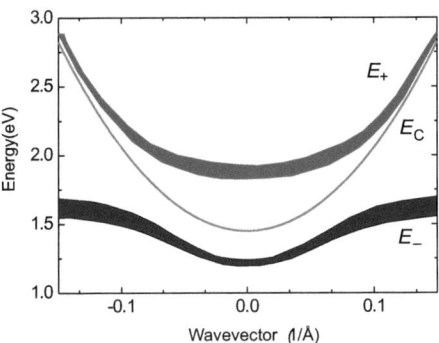

Figure 5.7: *The broadening of the conduction band dispersion curves shown using the example of the BAC model. The energy uncertainty is illustrated.*

$$N(E) = \frac{1}{\pi}\mathrm{Im} \int G(E, E_M) N_0(E_\mathrm{M}) \mathrm{d}E_\mathrm{M}$$

where N_0 is the unperturbed density of states of GaAs and E_M is the energy in the matrix semiconductor. The imaginary component of the Green's function can be calculated as following:

$$\mathrm{Im}\, G(E, E_\mathrm{M}) = \frac{\Omega(E)}{(\Gamma(E) - E_\mathrm{M})^2 + \Omega^2(E)},$$

where

$$\Omega(E) = -\sum_j \frac{V_j^2 x_j \Delta_j}{(E - E_j)^2 + \Delta_j^2}.$$

and

$$\Gamma(E) = E - E_\mathrm{C} - \sum_j \frac{V_j^2 x_j (E - E_j)}{(E - E_j)^2 + \Delta_j^2}.$$

Here, E_j are the nitrogen levels, Δ_j the nitrogen-related broadenings, and x_j the nitrogen concentrations. The V_j are the interaction energies giving rise to the hybridisation between the nitrogen states and the matrix semiconductor states, similar to the BAC model. E_C is the unperturbed conduction band edge of GaAs.

The effect of nitrogen is described by two impurity energy levels E_j, labeled E_N and E_NN. The first level represents single nitrogen atoms and the second one an averaged nitrogen pair. While the energy of the single nitrogen level is fixed relative to

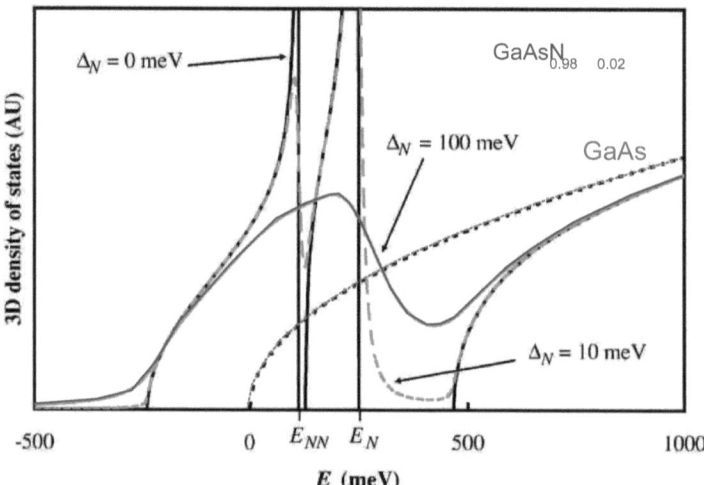

Figure 5.8: 3D DOS for GaAs$_{0.98}$N$_{0.02}$ calculated from the imaginary part of the Green's function using a three-band model for different nitrogen-related energy broadenings Δ_N [136]. The blue dotted line, starting at E=0 shows the GaAs conduction band DOS. The black solid lines, the gray dashed line, and the red solid line corresponds to the DOS, calculated for Δ_N=0 meV, Δ_N=10 meV and Δ_N=100 meV, respectively.

the conduction band minimum, the energy of the nitrogen pair states varies, depending on the nitrogen concentration.

Fig. 5.8 shows the calculated three-dimensional GaAs$_{0.98}$N$_{0.02}$ DOS, based on this model. The blue dotted line shows the GaAs DOS with the conduction band edge located at $E = 0$. The GaAsN DOS is calculated for different nitrogen-related energy broadenings Δ_N, where the broadenings on the two nitrogen-related energy levels are taken to be equal. In the limit of setting all the broadenings to zero, the DOS becomes infinite at the impurity energies E_N and E_{NN}, as can be seen in Fig. 5.8 by the black solid line. The introduction of a broadening Δ_N of 10 meV and 100 meV, as shown in Fig. 5.8 in the dashed gray line and red solid line, respectively, leads to an increasing smoothing of the DOS.

Comparing now this calculated DOS with the measured one, which is shown in Fig. 5.6(b), it can be seen that the characteristics of the calculated DOS for the broadening Δ_N of 100 meV (red solid line) is in excellent agreement with the measured DOS. The measured DOS confirms that nitrogen impurities change drastically the shape of the original GaAs DOS. Furthermore, the observed characteristics in the

5.5 Local density of states

conduction band confirms that the nitrogen incorporation leads to the strong scattering, which reduces the carrier mobility. Thus, the information gained using this three-band approach goes beyond that obtained from the usual BAC model.

Chapter 6
InAs(N)/GaAs(N) quantum dots

Self-assembled InAs/GaAs QDs have been intensively studied during the last years due to their promising technological applications [137, 138]. The goal to expand the emission wavelength of such QDs towards 1.3 µm or even to 1.55 µm is reached mainly by reducing the growth rate, embedding the QDs into an InGaAs QW [139, 140], or the stacking the QDs [119]. Another successful approach to reach 1.3 µm and 1.55 µm on GaAs substrates is the incorporation of nitrogen into InGaAs QWs and QDs [60, 141]. The motivation of nitrogen incorporation into the QD system is the expectation that nitrogen will reduce the lattice constant and thus the overall compressive strain in the sample, allowing the formation of larger QDs. At the same time, diluted amounts of nitrogen further reduce the optical transition energy. Thus, both effects lead to a red-shift of the photoluminescence (PL) wavelength.

Another method to expand the emission wavelength of InAs QDs towards 1.3 µm and even 1.55 µm can be the use of GaAsN as a capping layer or as an underlying layer for the QDs. A wavelength extension of more than 100 nm has been reported [142]. Such a red-shift can be related to the reduction of the matrix energy band gap and therewith of the electron confinement, determined by the difference in the QD states and the respective valence or conduction band of the surrounding matrix material [143]. However, a complete understanding is missing due to a lack of atomically resolved structural characterization.

6.1 InAs quantum dots within a nitrogen containing GaAs matrix

The PL wavelength of InAs QDs exhibits a strong blue-shift with increasing of the GaAs capping thickness [144]. Several processes are responsible for the observed blue-shift. First, during the capping additional strain within the QDs is introduced, caused by the lattice mismatch between the QDs and the capping matrix material. Second, during the capping different intermixing processes occur, changing the original stoichiometry and resulting in a less indium-rich concentration. Third, due to the capping the shape of the QDs is transformed from the pyramidal into the truncated pyramidal, considerably reducing the QD height and therewith changing the ground-state energy of the QDs.

However, the composition of the matrix material has an effect on the strain within the QDs and affects in this way the emission wavelength [145]. Moreover, the emission wavelength is determined by the electronic confinement. Increasing or decreasing the confinement leads to a blue-shift or red-shift of the emission wavelength, respectively. Finally, the structural parameters like size and shape of the QDs change with different surrounding matrix material and have an influence on the emission wavelength.

6.1.1 Sample structure and growth

In order to investigate structural changes induced by the nitrogen incorporated into the matrix surrounding InAs QDs, XSTM measurements were performed. Therefore, the sample was grown by O. Schumann [124] on a GaAs(001) substrate in a VG V80H solid-source MBE chamber. For the nitrogen incorporation again an Oxford applied research radio-frequency plasma source was used in the way, as discussed in section 5.1.

The sample contains stacked QD bilayers, which are separated by 15 nm of pure GaAs or $GaAs_{0.988}N_{0.012}$, as shown in Fig. 6.1. Thus, they should be stacked within a bilayer due to the local strain field [124, 146]. Figure 6.1 exhibits the sample structure as well as the corresponding XSTM overview images of the investigated sample. The QDs in the layers are marked by ellipses. The nitrogen plasma source was ignited after the growth of the first InAs/GaAs bilayer, therefore it does not contain any nitrogen.

The growth process was performed at a temperature of 490°C. Directly with the start of the InAs growth the arsenic pressure was reduced and kept at the reduced value until the end of the InAs growth. The nominal thickness of the deposited InAs material amounts to 0.7 nm (2.5 ML) for all grown QD layers. The material growth is further reported in Refs. [124, 125].

6.1 InAs quantum dots within a nitrogen containing GaAs matrix

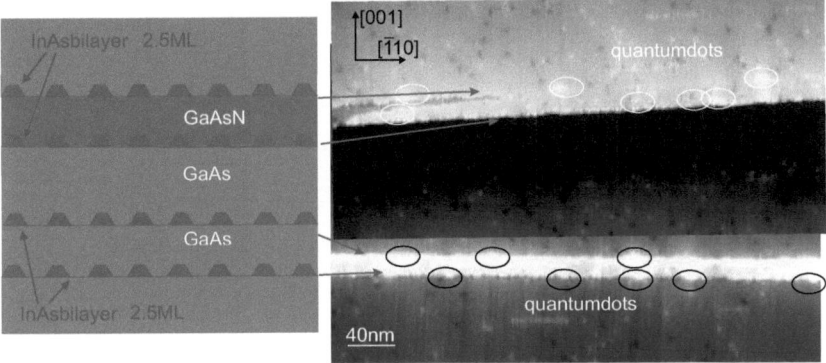

Figure 6.1: The sample structure (left) and XSTM overview images (right) of the 2.5 ML thick InAs/GaAs QD layers, acquired at –2.5 V and 70 pA. The QDs are marked by circles.

6.1.2 Structural changes induced by the nitrogen

In Fig. 6.2 XSTM images of InAs QDs grown on GaAs and on GaAs$_{0.988}$N$_{0.012}$ are presented. The InAs QDs appear bright compared with the GaAs host in these filled-state XSTM images due to both an increased tunneling probability by the QD states and the strain relaxation upon the cleavage [147]. The InAs QDs from the first layer completely embedded in GaAs [see Fig. 6.2(a)] have a truncated pyramidal shape with a base length extending to about $(20-30)$ nm along the $[\bar{1}10]$ direction and a height of $(3-4)$ nm along the [001] growth direction. Their density is determined to about 6×10^{10} cm^{-2}.

The InAs QDs grown on GaAs$_{0.988}$N$_{0.012}$ and capped with GaAs, exemplarily shown in Fig. 6.2(b), exhibit again a truncated pyramidal shape with a height of $(3-4)$ nm along the [001] growth direction, but a base length of only up to 20 nm along the $[\bar{1}10]$ direction. Thus, the resulting total volume of the InAs QDs grown on GaAsN is much smaller as compared with InAs QDs grown on GaAs. The density of these QDs is found to be 3×10^{10} cm^{-2}, which is decreased as compared with the one for the InAs/GaAs layer growth without any nitrogen.

In comparison with the InAs QDs grown on GaAsN, InAs QDs grown on GaAs and capped with GaAsN do not exhibit such a strong structural change. As it is shown in Fig. 6.2(c) their shape as well as their base length are similar to the InAs QDs completely embedded in the GaAs matrix. Further their density of 6×10^{10} cm^{-2} is found to be similar to the InAs/GaAs QDs grown without any nitrogen.

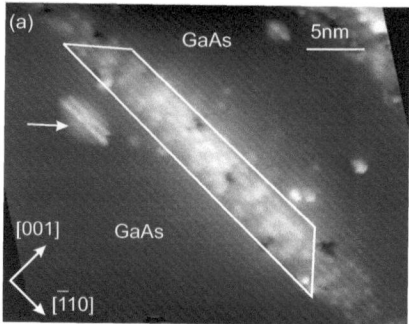

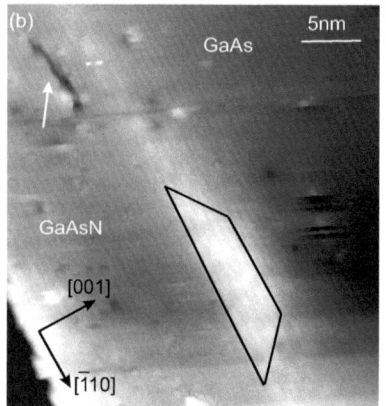
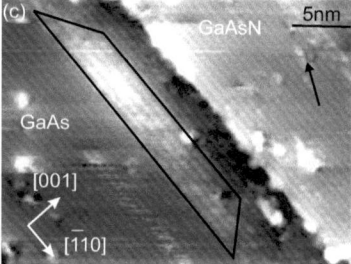

Figure 6.2: XSTM images of InAs QDs. The InAs QD in (a) is embedded in GaAs, the QD in (b) is grown on $GaAs_{0.988}N_{0.012}$ and capped with GaAs, and the QD in (c) is grown on GaAs and capped with $GaAs_{0.988}N_{0.012}$. The lines indicate the shape of the QDs. The XSTM images are taken (a) at –2.8 V and 90 pA, (b) at –3.3 V and 70 pA, and (c) at –2.2 V and 60 pA. The arrows indicate nitrogen-induced cleavage imperfections.

6.1.3 Stoichiometry determination

In Fig. 6.3 the stoichiometry of an InAs QD grown on and capped with pure GaAs [Fig. 6.3(a)] is compared with the stoichiometry of an InAs/GaAsN QD capped with GaAs [Fig. 6.3(b)] and furthermore with the stoichiometry of an InAs/GaAs QD capped with GaAsN [Fig. 6.3(c)]. Thereby the atomic chain distance across the center of the QDs (indicated by white dashed boxes) is derived from the XSTM images on

6.1 InAs quantum dots within a nitrogen containing GaAs matrix

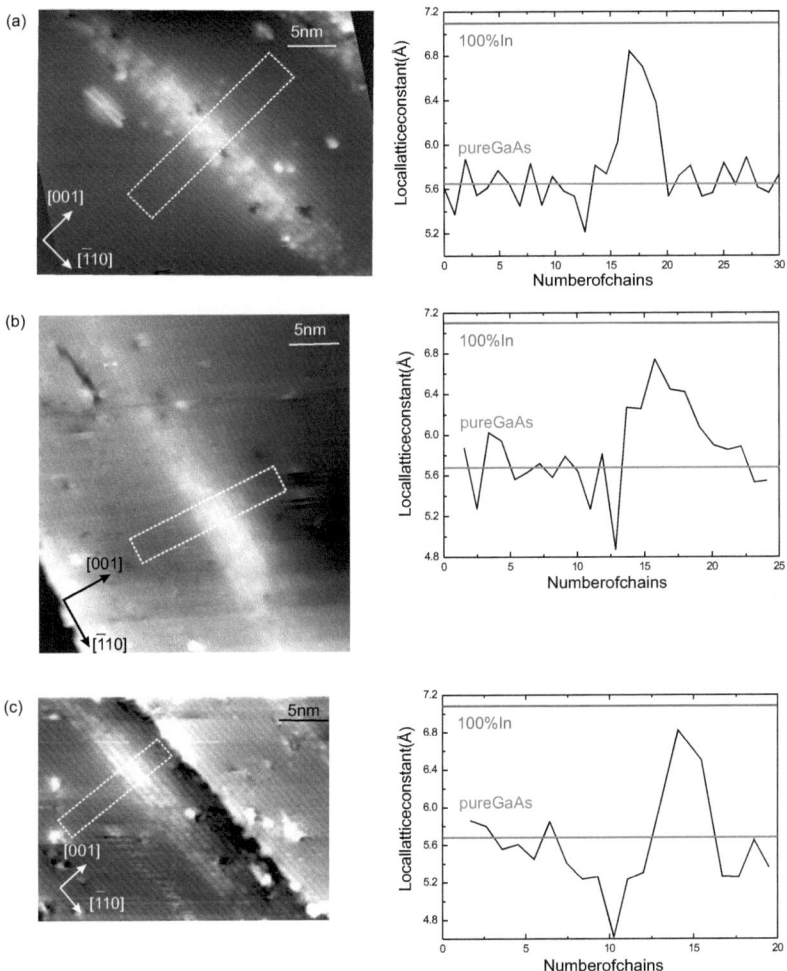

Figure 6.3: *Analysis of the chemical composition of three different InAs quantum dots: (a) grown on and capped with GaAs, (b) grown on GaAsN and capped with GaAs, and (c) grown on GaAs and capped with GaAsN.*

the left and plotted as function of the chain distance in the [001] growth direction on the right. This analysis method is discussed in detail in Ref. [71], where, moreover, the simulation of the strain relaxation for QDs with different InAs content is presented. The simulated lattice constant of a two-dimensional layer for InAs concentration close to 100% is indicated in Fig. 6.3 on the right side by solid red lines, whereas pure GaAs is marked by solid blue lines. In can be clearly seen, that all three QDs exhibit an maximal indium content of about 100%. Thus, the stoichiometry of the InAs QDs does not change, if moderate amounts of nitrogen are incorporated in the surrounding GaAs matrix.

6.1.4 Optical characterization

In order to study the optical properties of these QDs O. Schumann et al. performed PL experiments with a GaAsN layer only below or only above a single QD layer in order to distinguish between the influence of the strain and the influence of the confinement on the emission wavelength of the QDs [124]. The corresponding PL spectra are shown in Fig. 6.4.

The peak in the spectrum of each sample is assigned to the ground-state emission of the QD ensembles. There is a strong red-shift with increasing nitrogen content if the GaAsN layer is grown above the QDs, whereas the emission wavelength is practically not affected by the nitrogen for the samples with the GaAsN layer below the QDs. Moreover, the optical emission of these QDs shows a drastic degradation with increasing nitrogen content [121, 122]. This effect can be attributed to different phenomena, such as nitrogen being incorporated in interstitial positions [148] or compositional fluctuations in the alloy [149].

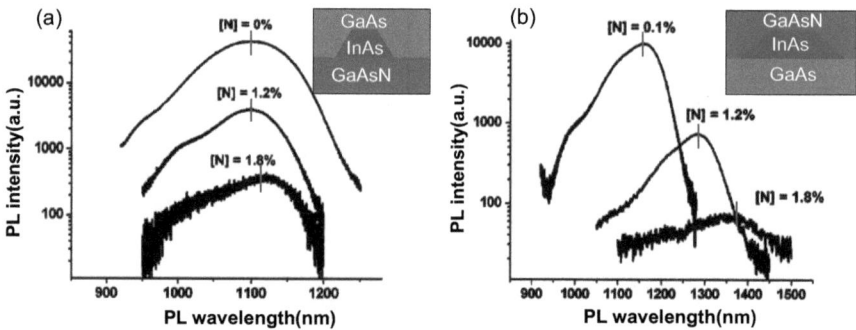

Figure 6.4: Photoluminescence spectra of the QDs at 300 K as a function of the nitrogen content for (a) InAs QDs grown on a 10 nm thick GaAsN layer, and (b) the QDs capped with a 10 nm thick GaAsN layer [124].

6.1.5 Discussion

The structural results show that InAs/GaAsN QDs demonstrate much smaller sizes, and thus a much smaller volume, as compared with InAs QDs grown on GaAs, while the emission wavelength of these QDs does not change significantly. In the case of InAs/GaAs QDs capped with GaAsN similar sizes were observed, as compared with InAs/GaAs QDs, capped with GaAs. But the PL wavelength of these QDs demonstrates a significant red-shift.

In order to understand this behavior O. Schumann et al. calculated the band structures for both InAs QDs grown on the GaAsN layer and capped with the GaAs and for InAs QDs grown on GaAs and capped with GaAsN [125]. In this calculation the conduction band and the valence band at the location of the QD exhibit almost no change. This indicates that the change in the strain inside the QDs due to the appearance of the GaAsN layer is small and does not affect the band structure. Thus, the change of the emission wavelength [see Fig. 6.4(b)] must come from the reduction of the confinement due to the nitrogen.

The emission wavelength of InAs QDs capped with GaAsN is extended beyond 1.3 μm, as compared with InAs QDs capped with GaAs. Because of equal sizes and strain, the reduction of the electron confinement has to be responsible for the observed red-shift. In contrast, the growth of InAs QDs on the GaAsN layer has almost no effect on the emission wavelength. This is due to a strong reduction in the volume of the InAs QDs grown on GaAsN. This volume reduction leads to a blue-shift of the emission wavelength and compensates the red-shift induced by the change of the confinement due to the nitrogen incorporation [124].

6.2 InAsN/GaAs quantum dots

Nitrogen can also be incorporated directly into InAs or InGaAs QDs to achieve a longer emission wavelength [60, 141]. Unfortunately, the photoluminescence (PL) intensity is again severely degraded at a high nitrogen content [60]. This degradation is usually attributed to a formation of deep levels such as bulk [150] or interfacial [151] traps, as observed in GaInAsN alloys grown on GaAs, or to the presence of dislocations within the InGaAsN QD layer [125]. Moreover, during the growth of InAsN/GaAs and GaInAsN/GaAs various alloying processes as well as segregation and surface diffusion occur, determining the final spatial and stoichiometric structure [152]. However, before this work the origin of the PL degradation has not yet been established and the exact spatial parameters as well as the shape of these QDs were unknown.

In this section the detailed structural and stoichiometric properties of such QDs are characterized.

6.2.1 Sample structure and growth

For the sample growth, again a VG V80H solid source molecular beam epitaxy chamber was used, which is assisted by an Oxford applied research radio-frequency nitrogen plasma source. The growth was performed at the constant sample temperature of 490°C, as described in section 5.1 and 6.1.1. The sample contains again QD bilayers separated by 15 nm of pure GaAs with nominal thickness of the deposited InAs and InAsN material of 0.7 nm (2.5 ML) for all grown QD layers, which should be stacked within a bilayer due to the local strain field [124, 146].

For this sample the nitrogen source was ignited before the growth of the QD layers. First, a reference bilayer was inserted to study the influence of the unavoidable background nitrogen pressure from the closed plasma source. It should be noted that the shutter of the burning nitrogen source cannot be completely tight so that GaAs layers grown with nitrogen plasma background usually contain about 0.1% of nitrogen [124], leading to an undesired nitrogen incorporation. During the growth, an advantage of this drawback was taken, keeping the nitrogen shutter closed all the time in order to offer very small amounts of nitrogen. While this nitrogen background concentration is estimated to be about 0.1% for the GaAs matrix, it is about 0.9% for the InAs QDs due to the lower growth rate for InAs.

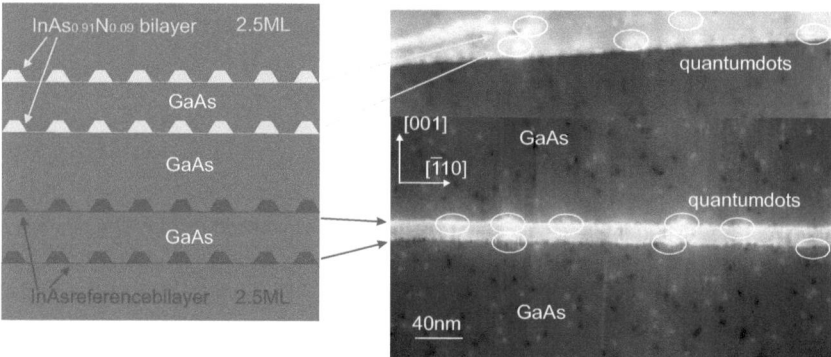

Figure 6.5: The sample structure (left) and respective XSTM overview images (right) of the 2.5 ML reference InAs/GaAs and the 2.5 ML nominal InAs$_{0.91}$N$_{0.09}$/GaAs QD layers, acquired at −2.5 V and 70 pA.

6.2 InAsN/GaAs quantum dots

Then, a second InAsN bilayer with nominally 9% nitrogen was deposited for QD growth within a GaAs host. The nitrogen concentration of this QD growth was varied again by changing the plasma parameters (flow of N_2) and further by changing the growth rate. The growth details are further reported in references [124, 125].

The sample structure and XSTM overview images of the reference InAs/GaAs layers and the nominal $InAs_{0.91}N_{0.09}$/GaAs layers are shown in Fig. 6.5. On the left side of this figure a schematic of the sample structure can be found. The arrows identify the different QD layers in the XSTM images. The reference bilayer and the nominal $InAs_{0.91}N_{0.09}$/GaAs bilayer are clearly visible as bright lines, containing bright spots, corresponding to single QDs, which are marked by ellipses in the XSTM overview images.

6.2.2 The reference bilayer

XSTM images of the reference bilayer show InAs/GaAs QDs with the typical spatial parameters for the applied growth conditions. Fig. 6.6 displays detailed XSTM data of QDs in the reference bilayer. The InAs QDs have a truncated pyramidal shape with a base length of about $(18 - 25)$ nm along the $[\bar{1}10]$ direction and a height of $(3 - 4)$ nm along the [001] growth direction.

For the stoichiometry determination [147, 153] along the colored boxes, marked in Fig. 6.7(a), the change in the lattice constant is observed and plotted in Fig. 6.7(b). The composition of this QD demonstrates an InAs-rich center with the maximum content of up to 100%.

The QD density is determined to about 7×10^{10} cm^{-2}, which is similar to density of the InAs QDs grown without any nitrogen, discussed in the section 6.2. An about 2 nm thick wetting layer (WL) [see Fig. 6.6(c)] is found between the QDs, which is intermixed due to segregation during capping. Its indium content reaches 20% at the base and exponentially decreases along growth direction.

The nitrogen content resulting from the background pressure of the closed plasma source can be directly determined from the filled-state XSTM images by counting the nitrogen atoms present at the anionic sublattice of the (110) cleavage surface [127]. Height profiles from the XSTM images allow to distinguish the nitrogen atoms from other crystal defects such as impurities, vacancies, or interstitials, as shown in chapter 5.4. For an accurate stoichiometry determination, an area of about 10,000 nm^2 was evaluated, resulting in a nitrogen content of about $(0.11 \pm 0.02)\%$ for the GaAs matrix, the WL, and the QDs. The observed nitrogen concentration for the GaAs is in agreement with the nominal one of 0.1%. Due to the different growth rates of InAs

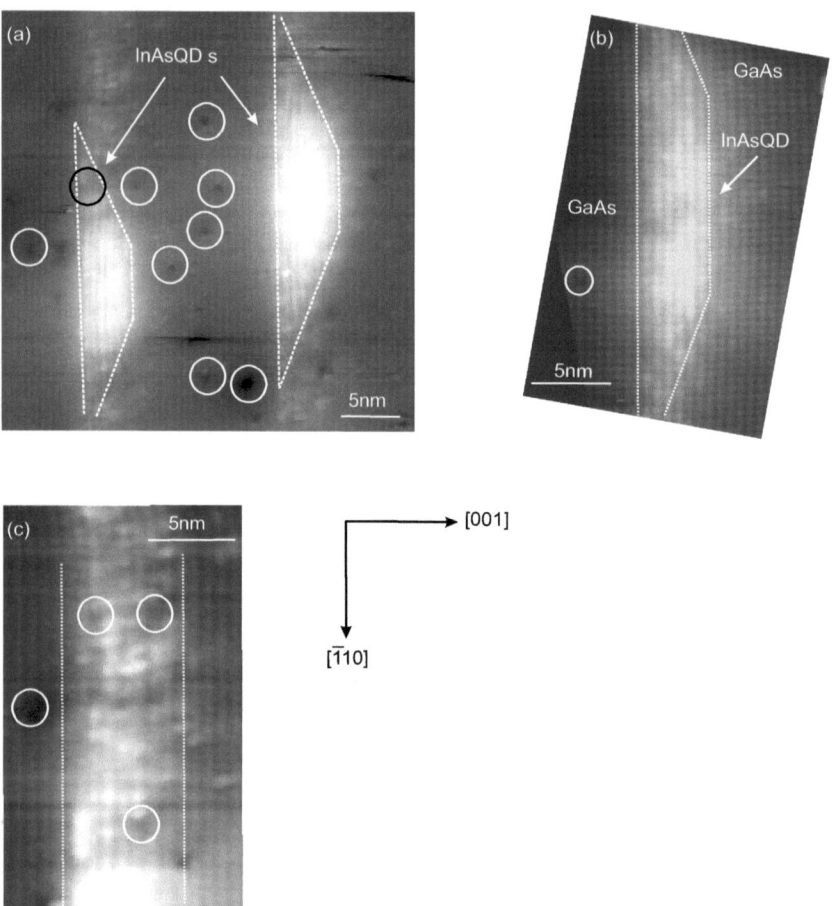

Figure 6.6: High resolution filled-state XSTM images of the reference bilayer, which only contains nitrogen from the background pressure. XSTM images of (a) two stacked InAs/GaAs QDs, (b) a single QD, and (c) a wetting layer, all acquired at –2.0 V and 70 pA. Locations of nitrogen atoms are marked by white circles in the GaAs matrix and the wetting layer and by a black one in the QDs.

and GaAs, the nominal nitrogen content from the closed plasma source is 0.9% in the InAs QDs of the reference bilayer. Thus, the actual nitrogen content in the QDs is about an order of magnitude lower than expected, already indicating a lower nitrogen incorporation into the InAs. This behavior will be discussed further below.

6.2 InAsN/GaAs quantum dots

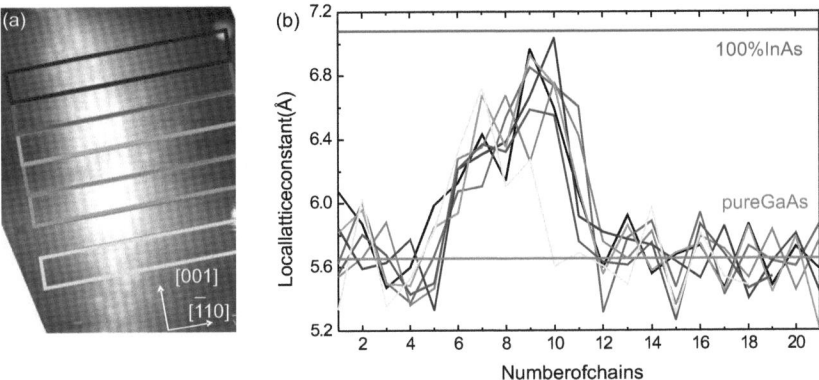

Figure 6.7: (a) XSTM image of a QD from the reference bilayer, taken at –2.0 V and 70 pA. (b) Results of the local stoichiometry evaluation [71]. The colored lines belong to the different stripes, marked by the correspondingly colored boxed in the image (a). The straight lines indicate the simulation results of the lattice constant for a definite In concentration. For the dark blue curve the In content reaches the maximum concentration of 100%.

In general, the XSTM observations show that the growth of the InAs/GaAs QDs is almost unaffected by this low nitrogen concentration of 0.1%, since their structural and stoichiometric properties are found to be very similar compared to usual QDs grown without burning nitrogen plasma [147].

6.2.3 The InAs$_{0.91}$N$_{0.09}$/GaAs bilayer

Compared with the QD growth in the reference layer, the nominal InAs$_{0.91}$N$_{0.09}$/GaAs QDs grown with the nitrogen shutter opened during InAs deposition show a completely different appearance in the XSTM images. Only their density of 6×10^{10} cm^{-2} is similar. As shown in Fig. 6.8, their size is dramatically increased, mostly along but also perpendicular to the [001] growth direction. The lateral extension varies from 20 to 30 nm along the $[\bar{1}10]$ direction, while the height reaches 8 to 10 nm along the [001] growth direction. The single QDs appear very different in their size, shape, and stoichiometry at the XSTM images. In general, instead of a truncated pyramidal shape more spherical shapes are found for all these QDs, as indicated by the dotted ellipse in Fig. 6.8(a).

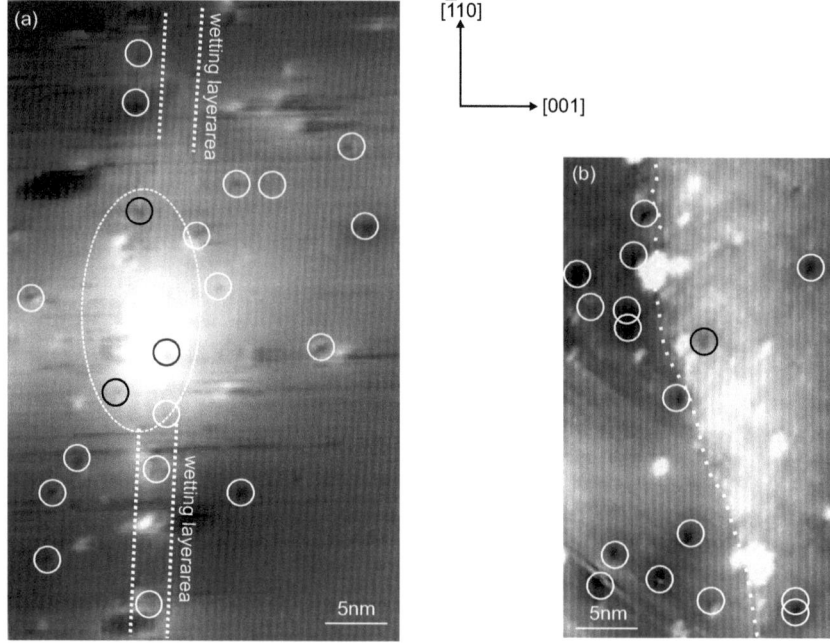

Figure 6.8: Filled-state XSTM images of two QDs from the nominal $InAs_{0.91}N_{0.09}/GaAs$ bilayer, taken at −2.5 V and 70 pA. Locations of nitrogen atoms are marked by white circles in the GaAs matrix and the WL region and by black ones within the InGaAs QDs. (a) The dotted ellipse indicates the QD shape and the dotted lines mark the area, where the WL should be present. (b) The dotted line indicates the interface between the QD and the underlying GaAs matrix material.

The composition of these QDs shows an maximum indium amount ranging from 15% to 40%, which is agian derived by analazing the local lattice constant [147, 153]. Figure 6.9(b) shows a representative example for the stoichiometry determination along the dotted box for the QD displayed in Fig. 6.9(a). This particular QD exhibits a maximum local InAs concentration of only 35%.

Moreover, indium containing material is found even underneath the nominal base plane where the QD growth was started, which can be clearly seen in Fig. 6.8(a). Such a behavior was up to now never observed for the case of the InAs/GaAs QD system. It strongly indicates that intermixing already occurs before the capping of the QDs, i.e. during the QD material deposition.

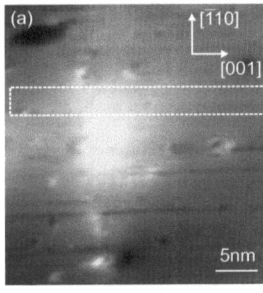 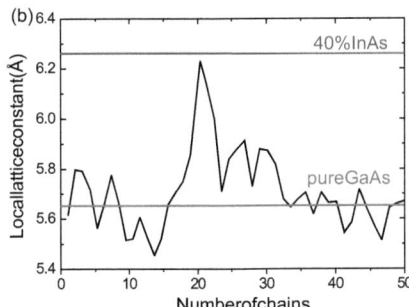

Figure 6.9: (a) Filled-state XSTM image of the nominal InAs$_{0.91}$N$_{0.09}$ QD layer, acquired at V_T=–2.5 V, showing a characteristic diluted InGaAs QD. (b) Results of the local stoichiometry evaluation. The solid line corresponds to the investigated stripe, indicated by the white dashed box in (a).

Nitrogen atoms are found in both the QD layer and the matrix material, as marked in Fig. 6.8. A nitrogen concentration of $(0.25 \pm 0.06)\%$ can be derived for the QDs, which is tremendously less than the nominal content of 9%. For the surrounding matrix material, in contrast, a nitrogen content of $(0.38 \pm 0.09)\%$ is found, which is definitely increased as compared to the nominal concentration from the background pressure of 0.1%. Thus, the nitrogen atoms are found to be segregated out of the indium-rich QD region into the neighboring GaAs matrix above and below the QDs.

Beside the diluted QDs, an incomplete, destroyed WL, or even no WL at all, is found. This can be seen in Fig. 6.10(a) in detail, where a uniform WL should be present between the white dashed lines, as expected for usual InAs/GaAs QD growth (e.g., in Fig. 6.6). The area, where the WL should be present is marked by dotted lines. It can be seen that almost no indium is present within this layer, and the position of the WL itself can only be estimated from the trace of remaining indium atoms and cleavage imperfections, the latter shown in Fig. 6.10(b). The dark depressions (trenches) as well as the bright protrusions (mesas, indicated by black ellipses) are cleavage imperfections typical for the nitrogen containing material [126], while a growth-induced formation of nanovoids [154] can be excluded because of the simultaneous observation of the mesas. These features are not only found at the WL but also above and below it, indicating further an increased amount of nitrogen also in the matrix material.

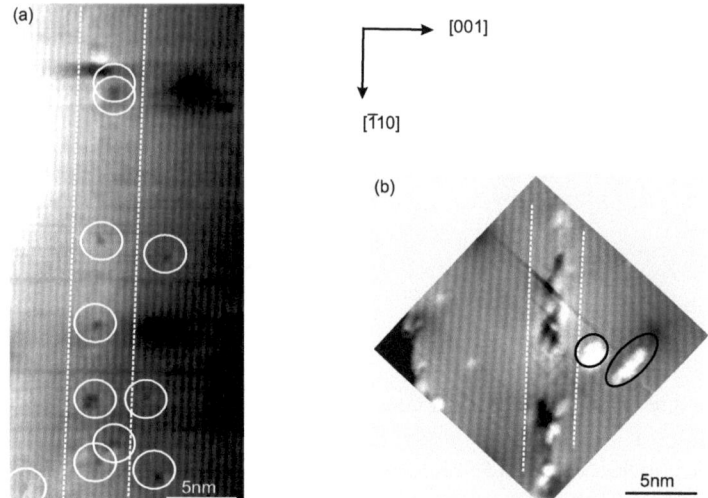

Figure 6.10: Filled-state XSTM images of the nominal InAs$_{0.91}$N$_{0.09}$/GaAs WL, taken at 70 pA and (a) –2.5 V and (b) at –2.2 V. Image (a) shows a completely missing WL, nitrogen atoms are marked by white circles. Image (b) shows nitrogen-induced cleavage defects. Nitrogen atoms are found mostly underneath and above the WL position. The remaining InGaAs does not form a closed layer. Black ellipses indicate the observed mesas.

Even if no direct evidence for dislocations of the crystal lattice is found in the XSTM images, they cannot be excluded completely, since they may be hidden by cleavage imperfections such as surface steps predominantly observed at the InGaAs containing regions. Dislocations may be generated due to the large lattice mismatch in these regions. In transmission electron microscopy images, dislocations were found to appear mainly in the WL, acting as nucleation centers for relaxed QDs [124]. The formation of dislocations would also be in good agreement with photoluminescence measurements, showing a strong decrease in intensity with increasing nitrogen content, as discussed in the next section.

6.2.4 Optical characterization

Upon incorporation of nitrogen into the QDs, a strong red-shift of the PL wavelength is expected, analogously to the incorporation of nitrogen into GaAs. In Fig. 6.11 it is shown that with increasing the nitrogen concentration the PL intensity at 16 K drops considerably. The peaks in the spectra of each sample are assigned to the

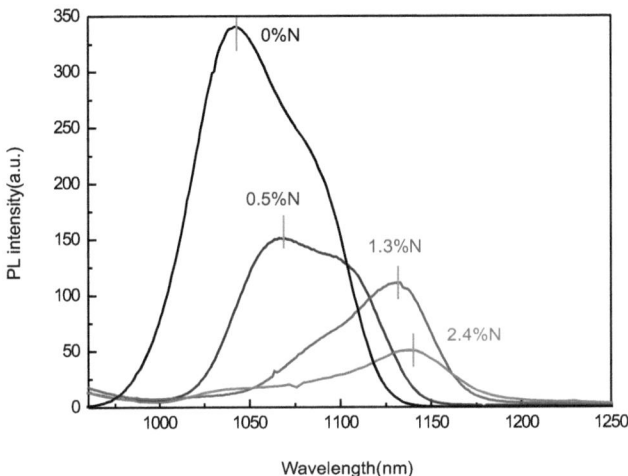

Figure 6.11: *PL spectra of samples with different nitrogen content, taken at 16 K. With increasing nitrogen concentration a red-shift of the PL wavelength occurs, accompanied by a drastical reduction of the PL intensity.*

ground-state emission of QD ensembles. The figure shows that with increasing of nitrogen content the wavelength demonstrates a moderate red-shift, and for higher nitrogen concentration the red-shift saturates. This indicates very strongly that the nitrogen incorporation into the QDs saturates for higher nitrogen concentrations or compensation effects due to a change of the QD structure take place.

6.2.5 Phase separation between indium and nitrogen

The segregation of the nitrogen atoms out of the InAsN material and the change of the shape of the QDs can be understood by taking into account that the solubility of nitrogen atoms in InAs is much lower than in GaAs [155, 156]. At the (001) growth surface and at the used temperature of about 490°C the arsenic-terminated $\beta 2(2 \times 4)$ reconstruction has the lowest surface energy. The surface energy of this reconstruction shows for GaAs and InAs similar values [157]. In contrast, the diluted $\beta 2(2 \times 4)$ InAsN surface has a much higher energy than the corresponding one of GaAsN [155]. Thus, the surface energy can be reduced after InAsN deposition and during capping by separating the nitrogen and indium atoms. However, this phase separation leads to a higher amount of strain due to the coexistence of indium-rich regions (QDs)

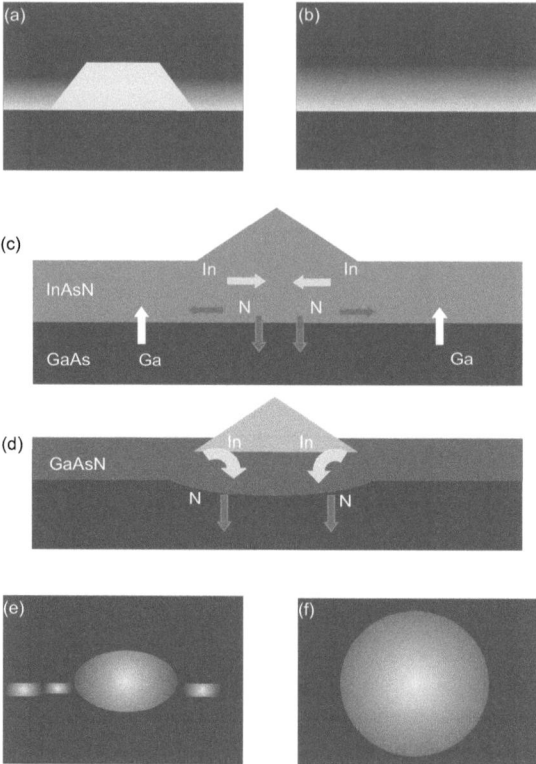

Figure 6.12: *Schematic model for the growth of InAs and InAsN QDs. (a) Usual InAs/GaAs QD characterized by a truncated pyramidal shape and a rather high indium concentration. The sharp interface between the InAs material and the underlying GaAs matrix strongly indicates that intermixing occurs only during capping by the matrix material but not during QD deposition. (b) The strong intermixing of InAs QD material with the capping GaAs material can lead to a laterally homogeneous two-dimensional layer in the case of sufficient segregation kinetics. This equilibrium structure also exhibits a sharp interface with the underlying GaAs matrix material. (c) The phase separation between indium-rich and nitrogen-rich regions and the resulting strain lead to (d) an intermixing of InAs material with the underlying GaAs matrix material and to (e) a destroyed WL and a diluted interface of the QDs with the underlying GaAs. (f) The absence of the WL can lead to the formation of a spherical InGaAs equilibrium structure in the case of sufficient segregation kinetics.*

and nitrogen-rich regions (matrix and WL area), since the lattice mismatch $\Delta a/a$

between pure InAs and pure (cubic) GaN amounts to 34% (see Tab. 2.1). This view is also supported by the observed surface roughness during the growth of GaInAsN layers [156, 158].

In usual InAs/GaAs QD growth, as observed in Fig. 6.6 and schematically shown in Fig. 6.12(a), the sharp lower interface strongly indicates that intermixing occurs only during capping by the matrix material [153], but not during QD material deposition. The tendency to form truncated pyramidal QDs during capping is related to the large strain at the QD apex evolving during capping as well as the slightly tensile strain along growth direction at the capped WL, both resulting in segregation processes from the QD apex to the side flanks. The final equilibrium structure along this segregation path would be a laterally homogeneous two-dimensional layer, as reported in Ref. [159] and schematically shown in Fig. 6.12(b).

In the case of the InAsN QD layer, in contrast, segregation processes are expected to be already strongly enhanced during QD formation due to the larger strain resulting from the phase separation into nitrogen-rich and indium-rich compounds [160]. This phase separation is schematically shown in Fig. 6.12(c). Due to this separation a strain-induced vertical intermixing of the InAs material with the GaAs underneath the QD occurs during growth, as schematically shown in Fig. 6.12(d), in agreement with the XSTM observations in Fig. 6.8. Furthermore, a stronger lateral indium segregation from the WL toward the QDs is also expected due to this phase separation, leading to a vanishing WL, at least in the case of a high QD density, as shown in Fig. 6.12(e).

As a consequence of the disrupted or missing WL, the final equilibrium structure comes closer to a spherical InGaAs inclusion in a GaAs matrix without any WL, which is demonstrated schematically in Fig. 6.12(f), and theoretically predicted [161]. This finding is in good agreement with the almost spherical shape observed here.

Chapter 7

Non-polar GaN surfaces

Group-III nitrides raised considerable attraction because of their ideal properties for green, blue, and ultraviolet laser and LED devices and for high-temperature electronics [2]. Therefore, intensive efforts have been invested in recent years to improve the quality of the epitaxial growth.

However, despite their significant success, conventional GaN-based laser diodes suffer from polarization-related electric fields that limit their optical efficiency. The deviation of the GaN unit cell from the ideal hexagonal wurtzite geometry and the strong ionic character of the group-III nitride bond are the origins of the polarization properties of wurtzite group-III nitrides. The polarization fields in wurtzite semiconductors are of two kinds: spontaneous polarization and piezoelectric polarization. Spontaneous polarization is caused by the deviation of the unit cell from the ideal hexagonal structure, thereby creating molecular dipoles in the material building up a macroscopic polarization field [162]. Due to the crystal symmetry the polarization is aligned along the [0001] direction. The piezoelectric polarization is caused by macroscopic strain-induced deformation of lattice constants, mainly presented due to a heteroepitaxial growth [162].

Figure 7.1(a) shows the $(1\bar{1}00)$ or m-plane of the wutzite structure and the schematic energy band diagram of an InGaN/GaN QW grown along the $[1\bar{1}00]$ direction without polarization effects. The main factors affecting the energies of the quantized levels are the potential wall height and the well width.

In Fig. 7.1(b) the (0001) or c-plane of the wutzite structure is shown. Furthermore, the directions and the effect of the spontaneous and piezoelectric polarizations on the band structure of InGaN/GaN QW grown along the [0001] direction are demonstrated [163]. In the InGaN QW the direction of the piezoelectric polarization P_{Pe} is opposite to the spontaneous polarization P_{Sp}, but both effects do not cancel, so

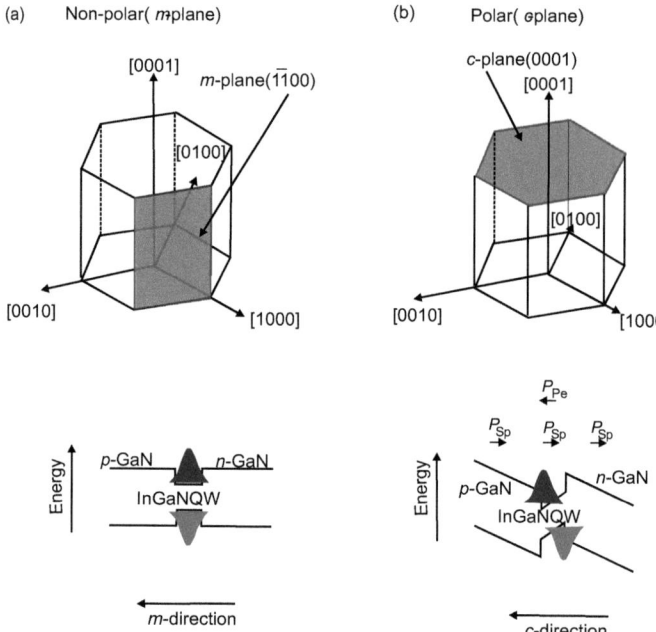

Figure 7.1: (a) Non-polar (m-plane) GaN. The schematic energy band diagram of an InGaN/GaN QW grown in m-direction shows an undistorted rectangular shape. (b) Polar c-plane GaN. The energy-band diagrams illustrate the separation of the electron (blue) and hole (green) wavefunctions for InGaN QWs [163].

that a triangular potential well shape results. Thus, the electron and the hole wave functions are shifted into the opposite sides of the QW and the energy difference of the electron and hole levels compared to the field-free structure is reduced.

Thus, laser diodes grown on c-plane GaN exhibit spontaneous piezoelectric polarization within the heterostructure [164]. These polarization effects generate electric fields and spatially separate the electron and hole wavefunctions, reducing their radiative recombination efficiency. For electrically injected laser diodes, extra carriers are required to screen these electric fields and flatten the distorted energy bands before efficient gain is provided. This process results in increased threshold current densities for the lasers [163].

To address these issues, structural growth can be performed on non-polar planes of GaN. These devices are free from the polarization-related electric fields, typical for c-plane structures [165]. The energy bands of InGaN QWs grown on m-plane GaN are undistorted, resembling the more rectangular shape of conventional QWs grown on GaAs or InP. These QWs do not suffer from the separation of the electron and hole wavefunctions found for c-plane structures. Additionally, with no polarization-related electric fields present, no extra carriers are required to achieve efficient optical gain. Indeed, higher optical gain is theoretically predicted for these structures [166].

For the growth of GaN substrates one particular challenge is the impurity, dopant, and defect incorporation, which often depends on the position of the Fermi level at the growth surface [167, 168]. Thus, it is of utmost importance to unravel the physical mechanisms governing the Fermi level position on group-III nitride surfaces. Indeed, for the most commonly used polar GaN(0001) growth surface, intrinsic surface states were suggested as origin of the Fermi level pinning [115, 169].

In contrast, for the non-polar GaN surfaces only little is known about the exact positions of the surface states and thus their possible influence on the Fermi energy. This is due to the lack of experimental data and disagreements between the existing theoretical calculations: Specifically, the reported energy positions of the surface states are inconclusive, as some calculations report that the empty dangling bond surface states lie within the fundamental bulk band gap [115, 170–172], while others found no surface states at all in the band gap [114]. It is, however, critical to know the exact energy position of surface states, because this allows to identify which physical mechanism governs the Fermi energy at the surface, such as intrinsic surface state pinning, surface defect pinning, or bulk doping for the case of unpinned surfaces.

7.1 The GaN($1\bar{1}00$) cleavage surface

In order to investigate the electronic structure of non-polar GaN surfaces and compare the results with theoretical calculations, several GaN(0001) oriented samples from different manufacturers were cleaved in ultrahigh vacuum along the $(1\bar{1}00)$ plane. On the GaN$(1\bar{1}00)$ cleavage surface the STM and STS experiments were performed. To ensure suffucient conductivity for the tunnel current n-type GaN with resistivities of $(8-10)$ mΩ were used. Different experimental results are presented in the following.

Figure 7.2(a) and (b) illustrate the typical morphology of the GaN$(1\bar{1}00)$ cleavage surface for two different cleavages. Both surfaces appear very similar in XSTM images, demonstrating that the GaN$(1\bar{1}00)$ cleavage surface consists of atomically flat terraces separated by monoatomic steps. The main difference between the cleavage surfaces shown here is their density of surface steps. The step densities are $(10^5 - 10^6)$ cm^{-1} and $(10^4 - 10^5)$ cm^{-1} for the cleavage surfaces in Fig. 7.2(a) and (b), respectively.

Figure 7.2: Constant-current empty state STM image of a cleaved GaN($1\bar{1}00$) surface (a) measured at +4.9 V and 80 pA, and (b) measured at +4.8 V and 80 pA. The terraces are separated by monoatomic high steps. (c) LEED pattern measured at 130 eV beam energy. The spot separation is consistent with a (1×1) surface structure. (d) Atomically resolved STM image showing the atomic rows of empty gallium dangling bonds along the [$11\bar{2}0$] direction, measured at +5 V and 90 pA.

Low energy electron diffraction patterns, shown in Fig. 7.2(c) and observed for both wafers, confirm a (1 × 1) surface unit cell, indicating that the surface is not reconstructed.

High resolution STM images, such as the one displayed in Fig. 7.2(d), show the atomic chains of the empty gallium dangling bonds along the $\left[11\bar{2}0\right]$ direction of the (1 × 1) ordered surface. This observed surface structure is in agreement with theoret-

7.2 STS investigations of the non-polar GaN($1\bar{1}00$) cleavage surface

ical expectations [114] and STM data of other cleaved wurtzite structure compound semiconductors [117], as presented in Fig. 3.10.

7.2 STS investigations of the non-polar GaN($1\bar{1}00$) cleavage surface

7.2.1 Density of states

Current-voltage spectra at different tip-sample separations are presented in Fig. 7.3. These data were taken on terraces of a cleavage surface with small terrace widths in the order of 10 nm, such a those shown in Fig. 7.2(a) and in the inset of Fig. 7.3. The spectra were measured sufficiently far away from surface steps, because of instable conditions during tunneling spectroscopy experiments on the surface steps.

At positive sample voltages all curves exhibit an onset of the current close to +1 V, while at negative voltages the apparent current onset strongly increases with

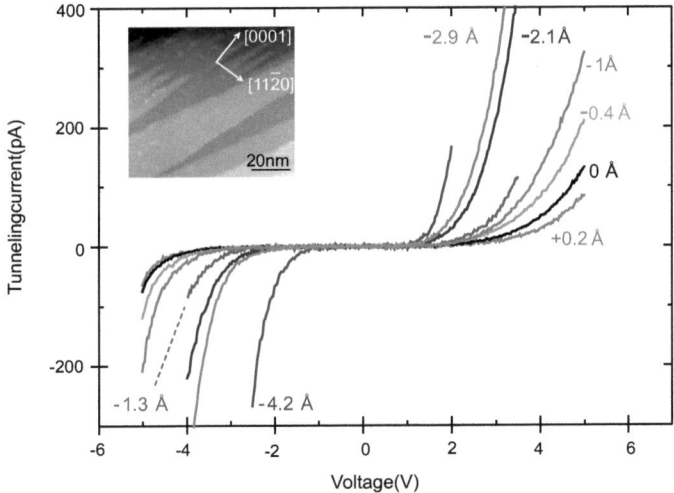

Figure 7.3: *Current-voltage spectra measured on narrow terraces of the cleaved GaN($1\bar{1}00$) surface. The different curves correspond to different tip-sample separations $z = z_0 + \Delta z$. Δz is given at each curve and z_0 is defined by the set voltage and current of +4.9 V and 80 pA, respectively. Inset: Constant-current empty state STM image of the terraces, where the spectroscopy data were taken, measured at +4.9 V and 80 pA.*

increasing tip-sample separation. As a result the apparent band gap changes with the tip-sample separation and is frequently even smaller than in bulk GaN. This raises the question, which mechanisms are responsible for this discrepancy, and where the band edges and the Fermi energy are actually positioned at the surface.

In order to identify the origins of the tunnel current and thus the positions of the band edges relative to the Fermi energy E_F, the logarithmic display of the current I and the normalized differential conductivity $(dI/dV)/\overline{(I/V)}$ as a function of the sample voltage are analyzed (Fig. 7.4).

The logarithmically displayed current curve in Fig. 7.4(a) exhibits (i) a clear onset at about $+1.0$ V of the tunnel current into the empty conduction band states of the surface (I_C) and (ii) two different current contributions at negative voltages, i.e. I_{acc} at voltages between 0 and -2.5 V, superposed by I_V at voltages below -2.5 V. Since the onsets of I_C and I_V are separated by about 3.4 V corresponding to the GaN band gap, the current contribution I_{acc} is related to energies within the band gap of GaN.

The different observed current contributions can be explained as follows:

(i) At positive voltages the tunnel current flows, if the Fermi level of the tip is energetically above the conduction band edge of the surface underneath the tip, as schematically shown in Fig. 7.4(c). Thus, at positive voltages the onset voltage at $+1.0$ V of the tunnel current corresponds to the position of the conduction band edge at the surface, E_C. This position of the conduction band edge indicates a Fermi level pinning at 1.0 eV below E_C.

(ii) Tunnel currents (such as I_{acc}) at voltages corresponding to energies within the fundamental band gap of a semiconductor were already observed previously on GaAs(110) surfaces [95, 102, 173]. The observation of such tunnel currents requires filled semiconductor states to face empty tip states. This can only occur, if (a) electrons accumulate at the GaN surface or if (b) intrinsic surface states exist in the band gap. As discussed below, filled intrinsic surface states do not exist in the band gap and thus case (b) can be ruled out.

One possibility of case (a) is that free charge carriers accumulate in the conduction band near the GaN surface due to a tip-induced band bending (TIBB), as discussed in section 3.2.3. This would lead to filled conduction band states, from which electrons can tunnel into the tip even at voltages corresponding to energies within the band gap [95, 102, 173]. However, in the present case, the surface is pinned 1 eV below E_C and thus TIBB would have to be larger than 1 eV in order to form an accumulation zone in the conduction band. Such large band bendings, however, do not occur at the used negative voltages smaller than -5 V: Even on unpinned GaN surfaces the band

7.2 STS investigations of the non-polar GaN($1\bar{1}00$) cleavage surface

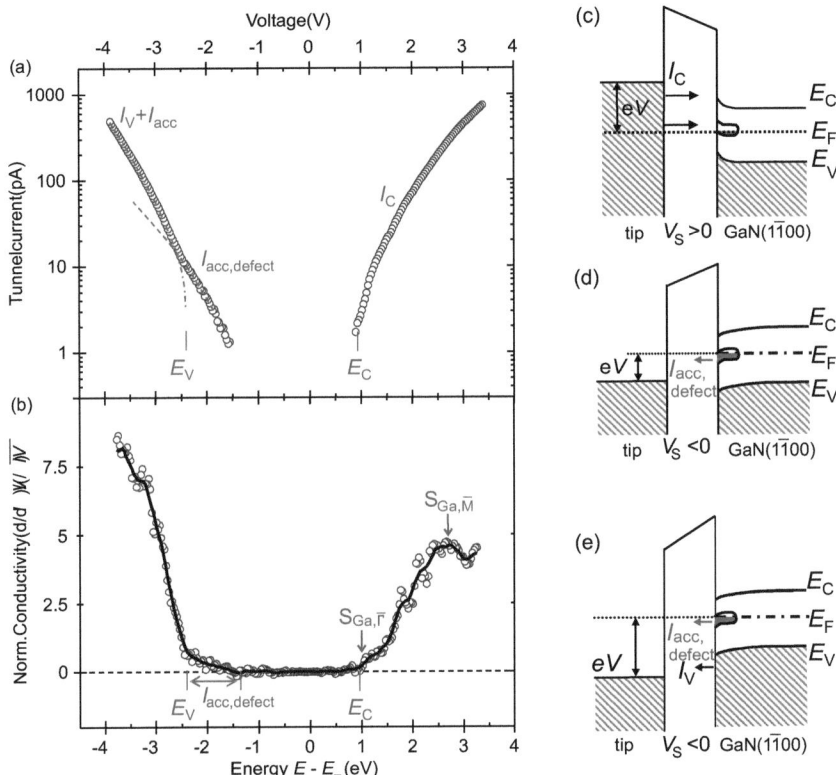

Figure 7.4: (a) Logarithmic display of the tunnel current as a function of voltage at a tip-sample separation of $z = z_0 - 3.3$ Å. (b) Normalized differential conductivity $(dI/dV)/\overline{(I/V)}$, derived from numerical differentiation. (c-e) Schematic diagrams of the origin of the (c) conduction band current, (d) the accumulation current, and (e) the valence band current superposed with the accumulation current.

bending calculated according to Refs. [95, 102] would be smaller than 0.75 eV. Thus, I_{acc} cannot arise from a tip-induced accumulation zone in the conduction band.

A further possibility of case (a) is the tip-induced accumulation of electrons in the states pinning the Fermi level (whose origin will be discussed below). These states lie directly at the Fermi energy and even a very small TIBB occurring on pinned surfaces can induce a considerable carrier accumulation in these states, as schematically shown in Fig. 7.4(d). Therefore, the accumulation current I_{acc} can be attributed to this effect.

(iii) If the magnitude of negative voltage is increased above the corresponding energy of the valence band edge E_V, also filled valence band states face empty tip states and additional electrons can tunnel from the GaN valence band into the tip yielding the I_V current contribution. This effect leads to a second onset of the tunnel current close to -2.5 V in Fig. 7.4(a), which thus corresponds to the valence band edge at the GaN (E_V), as schematically shown in Fig. 7.4(e).

The energetic positions of E_V and E_C can be seen even better in the normalized differential conductivity $(dI/dV)/\overline{(I/V)}$, shown in Fig. 7.4(b), which corresponds to the LDOS [92] (see section 3.1.3) and is normalized after [133], using $\overline{(I/V)}^2 = (I^2 + c^2)/V^2$ with $c = 70$ pA. At $(E_F - 2.4$ eV$)$ and at $(E_F + 1.0$ eV$)$ clear onsets of the valence and conduction band LDOS are visible, respectively. Thus, the band gap at the surface is (3.4 ± 0.2) eV wide, matching well the bulk band gap of GaN, supporting the identification of the band edges, and indicating the negligible effect of the TIBB on the derived energy positions due to the Fermi level pinning.

The minimum of the empty gallium dangling bond surface state (labeled $S_{Ga,\overline{\Gamma}}$) in Fig. 7.4) can be attributed to the small shoulder $(0.1 - 0.2)$ eV above the conduction band edge E_C [114]. Also the large peak at about $(E_F + 2.5$ eV$)$ ($S_{Ga,\overline{M}}$ in Fig. 7.4b) can also be assigned to this surface state. It can be related to the flat part of the band dispersion around the $\overline{M}$ point of the surface Brillouin zone (see Fig. 3.9) [114, 115], where the DOS is largest [174]. This interpretation is further supported by the measured inverse decay constant presented in the next section.

7.2.2 Intrinsic surface states dispersion

The dispersion of the intrinsic surface states can further be investigated by an estimation of the momentum of the tunneling electrons through an analysis of the exponential decay of the tunnel current $I \approx \exp(-2\kappa z)$ with increasing tip-sample separation z [175]. Fig. 7.5 shows the measured decay constant 2κ as a function of the voltage obtained for the largest tip-sample separations, where the effect of tip-sample interactions are negligible [176]. As shown in chapter 3, the decay constant 2κ can be approximated for one-dimensional systems and not too large voltages V [92, 102] by

$$2\kappa = 2\sqrt{\frac{2m_e}{\hbar^2}\left(B - \frac{|eV|}{2}\right) + \left|k_\parallel\right|^2}, \quad (7.1)$$

with m_e the electron mass and $k_\parallel$ the parallel wave vector of the tunneling electrons. In the absence of a reliable three dimensional barrier model, this approximation is used to illustrate the expected trend as a function of the voltage, which is shown as a solid line in Fig. 7.5 for $k_\parallel=0$ and an estimated effective tunneling barrier $B = 4.3$ eV.

7.2 STS investigations of the non-polar GaN(1$\bar{1}$00) cleavage surface

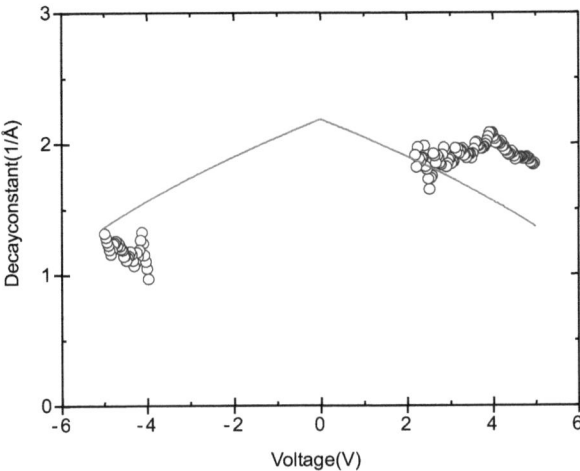

Figure 7.5: Decay constant 2κ as a function of the voltage (open circles). The solid line shows the calculated trend for tunneling only with zero parallel wave vector. These data contain information on the dispersion of the surface states.

At negative voltages and at small positive voltages (around 2 to 3 V) the measured values lie close to the calculated curve or even below, indicating tunneling of electrons with $k_\parallel=0$. Thus, the nitrogen and gallium derived dangling bond surface states form a direct band gap at the $\bar{\Gamma}$ point. At larger positive voltages the decay constant is larger than expected for tunneling with $k_\parallel=0$. This indicates tunneling with increasing parallel wave vector from states close to the edge of the surface Brillouin zone, in agreement with the predicted dispersion of the empty surface states (see Fig. 3.9) [114, 115, 170–172] and the above interpretation of the peaks in the conduction band.

7.2.3 Discussion

The accumulation current I_{acc} and its $(\mathrm{d}I/\mathrm{d}V)/\overline{(I/V)}$ signal are much smaller than those arising from the conduction and valence band states (Fig. 7.4). In addition, the current I_{acc} is more than four orders of magnitude smaller than the expected current from a tip-induced accumulation zone in the conduction band at the GaN surface, as calculated here according to Ref. [95] for a direct band gap at the $\bar{\Gamma}$ point. For this calculation, the determination of the positions of the conduction band and valence band edges at the GaN surface is based on the one-dimensional integration of the Poisson equation described by R.M. Feenstra and J.A. Strosio [102], but extended it in two significant aspects [177]. First, the Fermi energy in the bulk was calculated using Fermi-Dirac statistics. Furthermore, the equilibrium conditions for the carrier distribution in the semiconductor is not valid. Therefore, the effects of carrier dynamics zeroing the minority and/or majority carrier concentration at the surface are included, according to Ref. [95].

However, the filled DOS leading to I_{acc} is several orders of magnitude smaller than the one of an accumulation zone in the conduction band. This is in good agreement with an origin of I_{acc} in pinning states, whose concentration is typically in the range of 10^{11} to 10^{13} cm^{-2}, i.e. much less than the density of accumulated conduction band states or of dangling bond surface states, which are in the order of the surface states density of 10^{15} cm^{-2}. From these experimental observations it can be concluded that no intrinsic nitrogen and gallium derived dangling bond surface states are present in the fundamental band gap of GaN$(1\bar{1}00)$ surfaces and that the observed Fermi level pinning cannot arise from the dangling bond surface states. Instead it can be attributed to the large step densities of more than 10^6 cm^{-1}, corresponding to a step atom density of around 10^{14} cm^{-2} and to point defects observed on the GaN cleavage surfaces [174]. The contribution of the surface steps to the tunneling current is further limited, because the spectroscopy measurements were taken in a lateral distance to these steps.

Finally, it is noted that STS investigations were also performed on much broader terraces of cleavage surfaces, as shown in Fig. 7.2(b). In this case no stable spectroscopy data could be observed, in contrast to the above described STS measurements. Thereby, the only observed difference for both samples is the density of the surface steps. A high density of the extrinsic surface states, caused by a high step density of more than 10^6 cm^{-1} may be responsible for the strong Fermi level pinning discussed above, allowing a stable determination of the band gap. In contrast, the lower density of steps of about 10^4 cm^{-1} and the respective lower density of the extrinsic surface states may be not sufficient to produce the necessary Fermi level pinning.

7.3 Dislocations in GaN

GaN devices can only be produced successfully by epitaxial growth due to the high vapor pressure of nitrogen. Unfortunately, no well suited lattice- and thermally-matched substrates are available for the GaN epitaxy, so that GaN epitaxy leads to strain and huge dislocation densities [85, 86]. In order to reduce them, special techniques, such as epitaxial lateral overgrowth (ELOG), schematically shown in Fig. 7.6, or the inser-

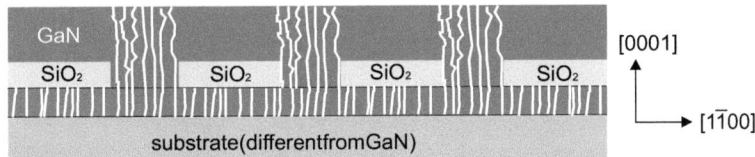

Figure 7.6: Epitaxial lateral overgrowth, used to concentrate the dislocations in bunches.

tion of strain reducing layers are used to concentrate the dislocations in bunches. In ELOG, as shown in Fig. 7.6 an oxide mask is deposited on the grown GaN film surface. The oxide mask is patterned and used to expose parts of the underlying GaN layer. GaN growth is continued, and laterally grown sections of GaN are free of dislocations, while other sections still contain high dislocation densities [178]. A major drawback of this method is that time consuming lithographic steps are required and that the samples need to be removed from the reactor. Despite such efforts, the dislocation density in commercial free-standing GaN wafers is still far above that of zincblende type III-V semiconductor substrates.

The presence of high dislocation densities is detrimental for optoelectronics, because dislocations can act as recombination centers. Therefore, large efforts focused on understanding the electronic properties of the different types of dislocations present in GaN. Unfortunately, thus far the experimental results are inconclusive: Dislocations in n-type GaN were found to be either all charged [179, 180], partially charged depending on the specific type [181–183], or uncharged [184]. In addition, strongly varying charge densities were reported for the same type of edge dislocation [179, 185]. In analogy to the experimental inconsistency, the charges of dislocations are attributed to the specific structure of the dislocation cores [114, 186–188], impurities and point defects [183, 184, 189, 190], or the introduction of strain-induced gap states [191]. One particular problem limiting the insight is that many experiments lack atomic resolution and/or are done on surfaces with surface states in the band gap. Therefore, dislocations in GaN were studied here using STM, enabling to probe simultaneously the charge state and structure of dislocations. Thereby two types of dislocations were identified, uncharged perfect and negatively charged dissociated dislocations.

7.3.1 Density of dislocations

Fig. 7.7(a) shows an overview of the $\mathrm{GaN}(1\bar{1}00)$ cleavage surface. The surface consists of atomically flat terraces separated by monoatomic steps. Most of the steps cross the whole image. However, four steps terminate suddenly at dislocations (see arrows). The

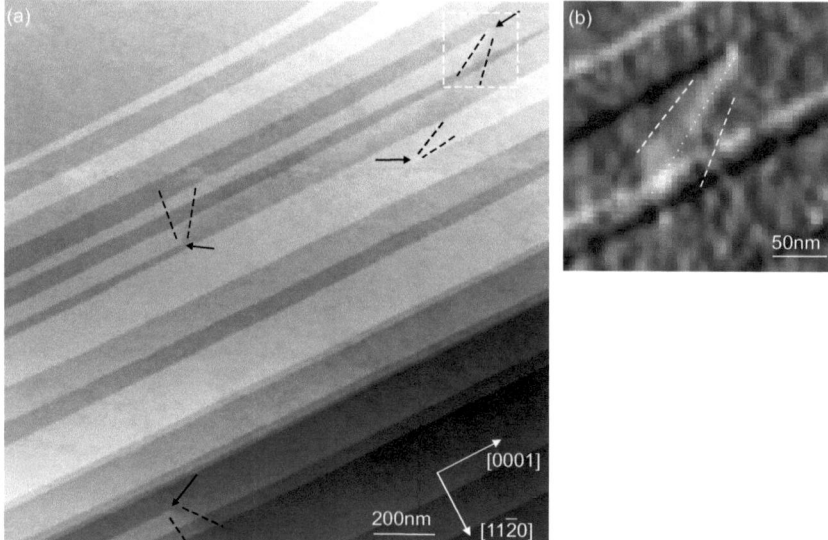

Figure 7.7: (a) Empty-state STM image of an area with four dislocations (see arrows) measured at +4.8 V and 82 pA. The vertical displacement fields of each dislocation are indicated by v-shaped dashed lines. (b) Gradient-filtered image emphasizing the vertical displacement field of the dislocation in the upper right corner of (a) indicated by a dashed square.

density of these dislocations was not homogeneous across the whole sample. In most areas no dislocations were found. Only in a localized region of about 5 μm diameter the dislocation density reached values of 1.2×10^8 cm^{-2}. The spatial localization of dislocations indicates that the dislocations form bunches, while the remaining material is free of dislocations. For comparison the nominal threading dislocation density of the investigated GaN material is $(1-2) \times 10^7$ cm^{-2}.

7.3.2 Determination of the line direction from STM images

During growth of GaN layers on non lattice-matched substrates, dislocations form near the substrate-GaN interface. Their line directions, initially primarily following the [0001] growth direction [192, 193], may bend with progressing growth toward the side facets [192], as shown in Fig. 7.8. Thereby the dislocation lines may cross the $(1\bar{1}00)$ cleavage surface investigated here. Since dislocations were observed on the cleavage surface, their lines must have indeed bent away from the [0001] direction.

7.3 Dislocations in GaN

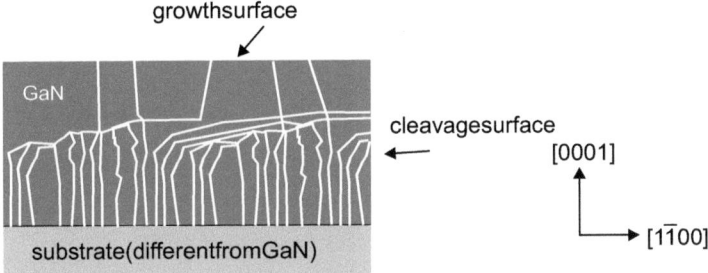

Figure 7.8: *Creation and development of dislocations at the substrate-GaN interface. Due to special growth conditions the dislocations bend away from the [0001] direction. Some of them cross the $(1\bar{1}00)$ cleavage surface.*

In fact, the component of the dislocation line direction projected on the surface can be extracted from the STM images. Each dislocation induces a displacement field, which is largest close to the dislocation core and decays with increasing distance. If a dislocation line crosses the cleavage surface non-perpendicularly, the displacement field at the surface is largest where the subsurface dislocation core is closest to the surface. The imaged horizontal gradient $(\mathrm{d}z/\mathrm{d}x)$ [Fig. 7.7(b)] illustrates the v-shaped vertical displacement field of one dislocation. The largest displacements occur along the dotted line. Thus, the dislocation line has a projected component along the same direction. The diagonal contrast lines in Fig. 7.7(b) arise from steps and are not of interest. Analyzing each dislocation in this manner yields that every dislocation has a different line direction [see v-shaped dashed lines in Fig. 7.7(a)]. This indicates that the dislocations form bunches of entangled nonparallel dislocation lines.

In the STM images, two types of dislocations were found [194], which are presented in the following.

7.3.3 Perfect dislocation

Figure 7.9 exemplifies a perfect dislocation. The STM image shows the empty dangling bonds above the gallium atoms [117]. The dislocation core is located at the end of the surface step (arrow). Note that the step makes a sharp turn toward the $\left[11\bar{2}0\right]$ direction 5 nm before the dislocation core.

The displacement field (v-shaped dashed lines) suggests that the dislocation line is approximately lying in a (0001) plane. The three components of the Burgers vector were determined as follows. First, the dislocation induces a step with a height of 1 ML at the $\left(1\bar{1}00\right)$ surface. Thus, the Burgers vector has a component of $a/2\left[1\bar{1}00\right]$

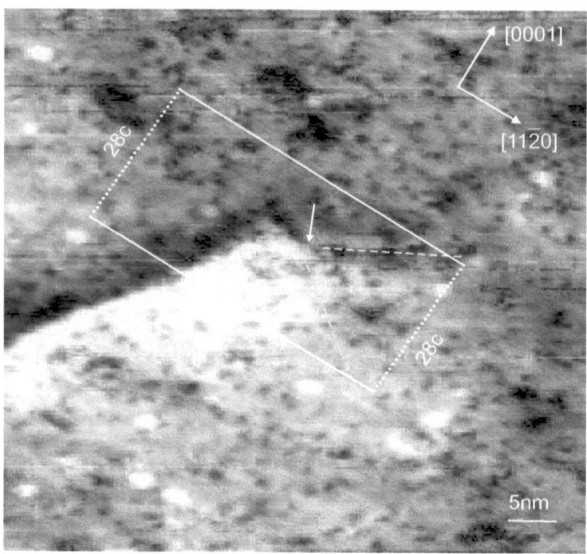

Figure 7.9: Empty-state STM image of a perfect dislocation with a Burgers vector of type $a/3\langle 11\bar{2}0\rangle$ measured at +4.2 V and 86 pA. The vertical displacement field is indicated by the v-shaped dashed lines. The solid lines and points are used to illustrate that the Burgers vector has no component along the [0001] direction.

perpendicular to the surface. Second, the separation of the two atomic rows in the $[11\bar{2}0]$ direction marked by solid white lines equals 28 c on both sides of the dislocation core, as indicated by the white points. Thus, the Burgers vector has no component in the (0001) direction. Finally, the step induces not only a component of the Burgers vector perpendicular to the surface, but also parallel to the atomic rows within the surface plane, because two neighboring $(1\bar{1}00)$ planes in the wurtzite structure are offset by $a/6\left[11\bar{2}0\right]$. This yields a total Burgers vector of $a/2\left[1\bar{1}00\right]\pm a/6\left[11\bar{2}0\right]$, i.e. of the general type $a/3\left\langle 11\bar{2}0\right\rangle$, tilted 30° with respect to the surface normal. This is indeed the only possible Burgers vector without any component along the (0001) direction in wurtzite crystals [79].

7.3.4 Dissociated dislocation

Fig. 7.10(a) shows the second observed type of dislocation, which is dissociated into two partial dislocations marked P_1 and P_2. The first partial dislocation core (P_1) is located at the point, where the monoatomic step at the $(1\bar{1}00)$ surface (1 ML ≈ 2.75 Å)

7.3 Dislocations in GaN

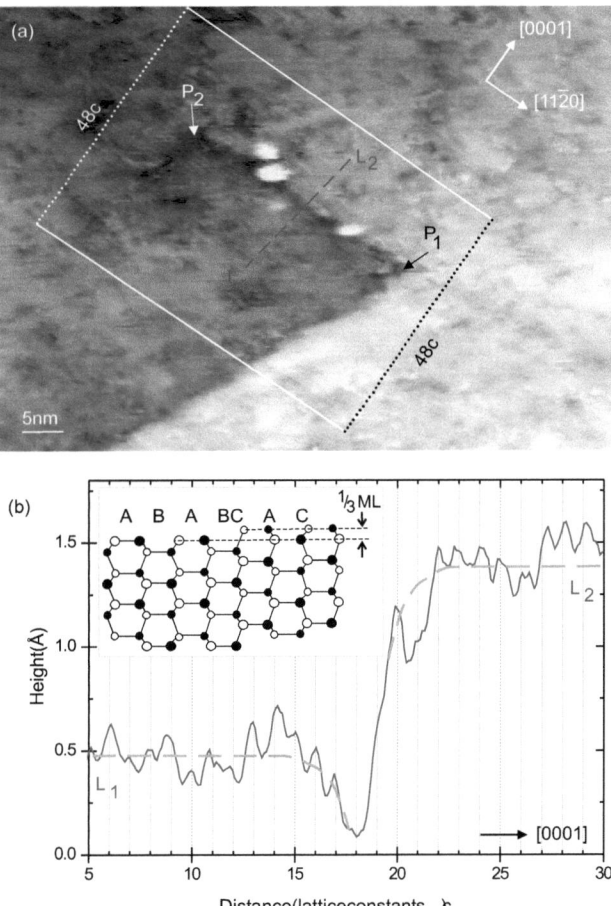

Figure 7.10: (a) Empty-state STM image of a dislocation with an overall Burgers vector of type $a/3 \langle 11\bar{2}0 \rangle$ dissociated into Shockley partial dislocations P_1 and P_2 (measured at +4.0 V and 94 pA). The solid lines and points are used to illustrate that the Burgers vectors has no component along the [0001] direction. (b) Height profile measured along the dashed line L_1-L_2 in (a). The stacking fault between the two partial dislocations results in a 1/3 ML step. Inset: schematic of an intrinsic type-2 stacking fault. Black and white circles represent gallium and nitrogen atoms, respectively.

sharply turns and transforms into a 1/3 ML high step (height ≈ 0.9 Å). The second partial dislocation is positioned at the end of the 1/3 ML step (P_2). Between both partial dislocations a stacking fault in the (0001) plane is present, which results in a 1/3 ML height shift visible in the height profile shown in Fig. 7.10(b).

The Burgers vectors are determined as above. The two atomic rows in the $\left[11\bar{2}0\right]$ direction on both sides of the dissociated dislocation marked by white lines are separated everywhere by 48 c. Thus, the total and partial Burgers vectors have no component in the [0001] direction. The monoatomic step yields again a Burgers vector of the type $a/3\left\langle 11\bar{2}0\right\rangle$. Dislocations with such Burgers vectors can dissociate into two Shockley partial dislocations with Burgers vectors of the type $a/3\left\langle 1\bar{1}00\right\rangle$ [80], namely $a/3\left[2\bar{1}\bar{1}0\right] = a/3[1\bar{1}00] + a/3[10\bar{1}0]$ or $a/3[1\bar{2}10] = a/3[1\bar{1}00] + a/3[0\bar{1}10]$.

With this in mind, the reduction in the step height from 1 to 1/3 ML at the partial dislocation P_1 corresponds to a Burgers vector of $a/3\left[1\bar{1}00\right]$ oriented parallel to the surface normal. The partial dislocation P_2, where the 1/3 ML height shift disappears, has the same type of Burgers vector, but rotated 60° away from the surface normal.

This identification agrees with the structure of the stacking fault. As discussed in section 2.5.2 and shown in Fig. 2.9, three different types of stacking faults can occur in the wurtzite structure. Two of these stacking faults cause the observed height difference of 1/3 ML. However, these two stacking faults are not connected with the same row pattern. The stacking fault of type $\{ABABCACAC\ldots\}$ causes no changes in the row spacing near the partial step edge [Fig. 2.9(c)], while the stacking fault of type $\{ABABCBCB\ldots\}$ yields one half row spacing near the edge [Fig. 2.9(b)].

The height profile in Fig. 7.10(b) shows not only the required height offset of 1/3 ML, but also the absence of any shift along the [0001] direction of the $\left[11\bar{2}0\right]$ oriented atomic rows, because the positions of the height maxima on both sides of the stacking fault exactly match the dotted lines marking the lattice. Thus, this structure corresponds to a double fault with a stacking sequence $\{ABABCACA\ldots\}$ (type-2 intrinsic stacking fault), as shown in Fig. 2.9(c) and further in the inset in Fig. 7.10(b). Such a stacking fault is connected to Shockley partial dislocations [80], supporting the above identification of the Burgers vectors and dislocation types [194].

7.3.5 Charge of dislocations

A close look at the height profile of the stacking fault shows an apparent height depression close to the stacking fault, as indicated by the gray dashed lines in Fig. 7.10(b). Similar depression zones were observed previously around charged point defects and steps [195–197], where the depression arises from an upward band bending induced by

7.3 Dislocations in GaN

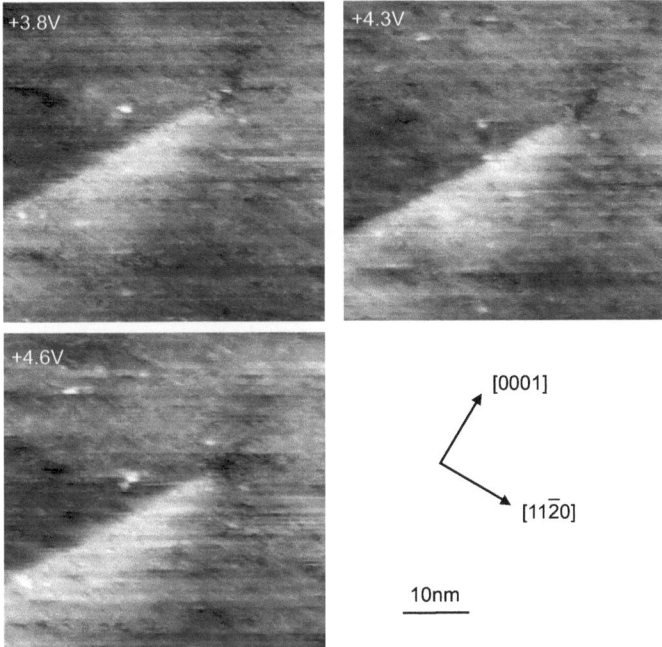

Figure 7.11: STM images of a perfect dislocation at different applied bias voltages.

localized negative charges. The band bending reduces the density of empty states in the conduction band available for tunneling at positive voltages [195–197]. Therefore, the tip is moved closer to the sample in order to keep the tunnel current constant. This leads to the observed depression zone. Thus, the stacking fault is negatively charged. Similarly, around the partial dislocation cores dark zones are found, suggesting a negative charge also of the partial dislocations. In contrast, no such dark zones were found at the perfect dislocations, e.g. in Fig. 7.9 and Fig. 7.11.

Fig. 7.11 shows another perfect dislocation at different tunneling voltages. In both Fig. 7.9 and Fig. 7.11 the surface exhibits a number of charged point defects surrounded by a dark contrast, while no such dark contrast is visible at the dislocation core itself. By applying a different bias voltage this appearance does not change, as it is shown in Fig. 7.11. Thus, perfect dislocations are uncharged on n-type GaN. These observations indicate that the charge transfer level $(0/-1)$, i.e. the energy where the charging switches from neutral to negative, is lower in energy for the partial than for the perfect dislocation.

7.3.6 Discussion

The charges of the partial dislocations are unlikely to be induced by charged impurities/defects attracted in the strain field or strain-induced deep electronic states, because the strain field is reduced for partial dislocations by dissociation as compared to perfect dislocations [114, 186–188]. Thus, the different charge states are rather originating in the specific core structure of the different dislocations. Indeed, Shockley partial dislocations were predicted to have deep levels in n-type GaN [198]. The charge of the stacking fault itself might arise from additional surface dangling bonds at the edge of the 1/3 ML height shift. However, the stacking fault is a local insertion of cubic GaN with the stacking sequence (ABCA) into the wurtzite structure. Since cubic GaN has a smaller band gap than wurtzite GaN (see Tab. 2.1), the stacking fault may introduce localized gap states. It should be noted that a charging of Shockley partial dislocations and the corresponding stacking fault was also observed in n-type GaAs by STM [199].

7.4 Doping modulation in GaN

Epitaxial layers of group-III nitrides developed rapidly towards the system of choice for green to ultraviolet optoelectronic devices. Such devices are based primarily on heterostructures of different group-III nitrides and/or differently doped layers [20, 200, 201], where the exact atomic arrangement within the epitaxial layers of the nanostructures and at the interfaces sensitively influences the optoelectronic properties [202].

XSTM is a particularly attractive tool for the high resolution analysis of interfaces and epitaxial layers in compound semiconductors [203, 204]. Indeed, for zincblende type materials, XSTM provides a direct atomically resolved access to structural and electronic bulk properties [134, 159, 204–209], thereby helping to unravel the processes during epitaxy, as illustrated in the preceding chapters. This technique is in principle also applicable to wurtzite structure semiconductors due to the presence of suitable cleavage surfaces [117]. However, before this work no epitaxial structure or interfaces in group III-nitrides could be successfully imaged by XSTM. In this section a doping modulation in GaN is investigated by XSTM and the contrast mechanisms are identified.

7.4 Doping modulation in GaN

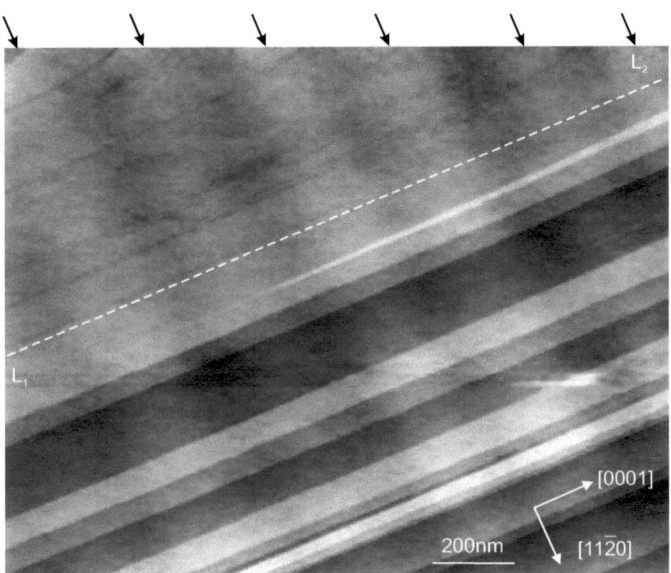

Figure 7.12: *Cross-sectional empty state constant-current STM image showing a doping-induced contrast modulation along the [0001] direction. The height maxima are marked by arrows. The steps, oriented primarily along the [0001] direction, do not affect the contrast modulation. The image was measured at +5.9 V and 80 pA.*

7.4.1 STM investigations

Fig. 7.12 shows an overview image of the $\text{GaN}(1\bar{1}00)$ cleavage surface of an epitaxial GaN layer. The surface consists again of atomically flat terraces separated by mono- and diatomic steps. These steps are oriented primarily along the [0001] direction and are mostly cleavage related.

In addition, Fig. 7.12 shows a contrast modulation along the [0001] direction with about six periods visible in this particular image. The maxima of the contrast modulation are marked by arrows. This contrast modulation is not influenced by any surface defect, e.g. steps. In particular, it is found to be independent of the step density. The average separation between adjacent maxima of this contrast modulation is (270 ± 30) nm.

In order to analyze the contrast modulation, height profiles along the [0001] direction are extracted from an atomically flat terrace, as shown in Fig. 7.13(a). All height profiles acquired for different voltages using the same tip (labeled tip 1) exhibit very similar height modulations. However, at a closer look, the modulation amplitude is

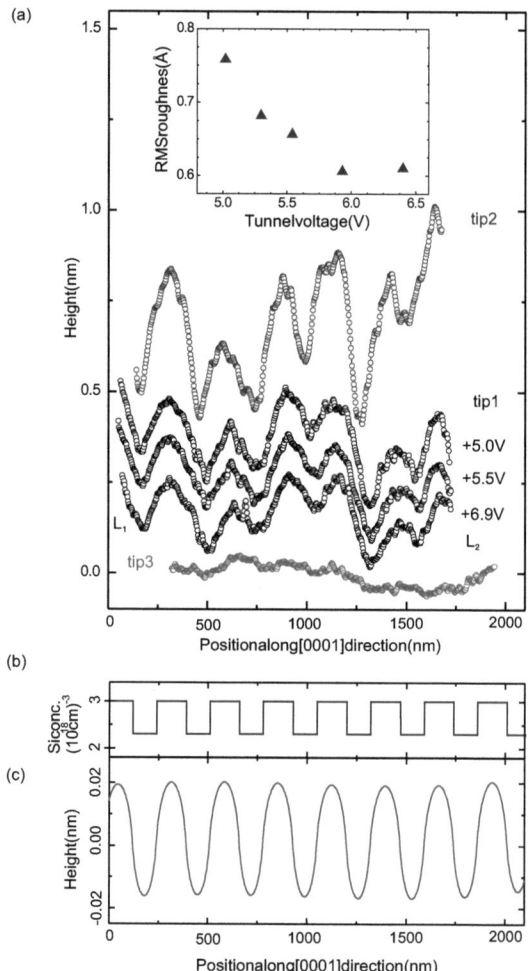

Figure 7.13: (a) Height profiles of the contrast modulation along the dashed line in Fig. 7.12. The profiles were acquired at different voltages and with three different tip configurations. Inset: Root-mean square roughness of the height profiles for different tunneling voltages acquired with tip 1. (b) Assumed silicon concentration profile used to simulate (c) the height modulation due to the strain relaxation at the GaN cleavage surface.

7.4 Doping modulation in GaN

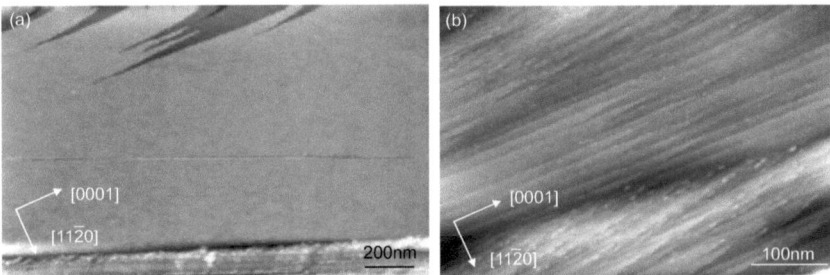

Figure 7.14: *Cross-sectional empty state constant-current STM images of the GaN[1̄100] surface without any doping-induced contrast modulation. The absence of this contrast is independent of the step density, which is low in (a) and high in (b). The steps are primarily orientated along the [0001] direction. Both images are measured at +4.9 V and 80 pA.*

decreasing slightly with the applied voltage, as illustrated by the corresponding root mean square (RMS) roughness as a function of voltage in the inset of Fig. 7.13(a). In contrast, for different tip configurations obtained by tip instabilities, occurring during repeated scanning of the same location, the amplitude is changing drastically. For example, tip 2 yields a very large modulation height (corresponding RMS roughness is 1.4 Å). In contrast, tip 3 yields a very small modulation height with a corresponding RMS roughness of only 0.3 Å, while the actual value related to the modulation is even smaller because of a slowly varying background and the noise level. In the latter case the height between the neighboring maximum and minimum amounts to about 0.4 Å. However, a contrast modulation is observed with all tips: The smallest observed modulation height was found to be close to 0.4 Å, as shown, e.g., for tip 3. Note, that the positions of the modulation maxima and minima do not shift with different tunneling voltages or tips.

For another epitaxial GaN layer no comparable contrast modulation is observed in XSTM images, as shown in Fig. 7.14. The absence of this contrast is independent both of the step density, which is low in Fig. 7.14(a) and high in Fig. 7.14(b), and of the applied bias.

7.4.2 Secondary ion mass spectroscopy investigations

In order to unravel the origin of the observed contrast modulation, the spatial distribution of different elements along the [0001] direction were measured by secondary ion mass spectroscopy (SIMS). For the epitaxial layers presented in Fig. 7.12 and Fig. 7.13 all probed elements except silicon were found to be constant. Fig. 7.15 shows that

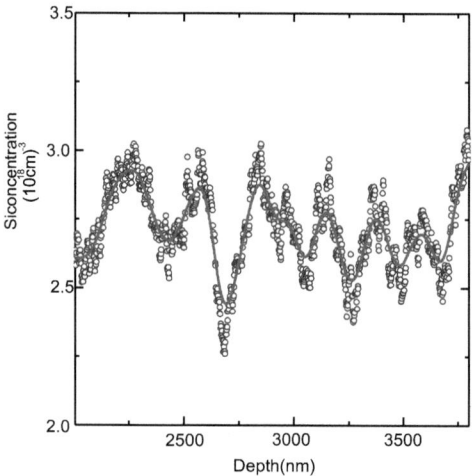

Figure 7.15: *Silicon concentration in GaN along the [0001] direction as measured by SIMS (blue circles). The red line is a smooth for guiding the eye.*

the silicon concentration exhibits a significant modulation. The concentration varies between 2.3×10^{18} and 3.0×10^{18} cm^{-3}. The maxima of the modulation of the silicon concentration have an average separation of (270 ± 20) nm. For the second epitaxial layer imaged in Fig. 7.14, in contrast, the silicon concentration does not exhibit any modulation, confirming the corresponding XSTM results shown in Fig. 7.14.

7.4.3 Origin of the observed modulation

First of all, a different origin of the contrast modulation, standing electron waves, will be discussed. The observed wavelength of 270 nm, however, would be extremely long as compared to previous observations of standing waves on, e.g., Cu(111) [210] or GaAs(110) [211] surfaces. Such a long wavelength would correspond to extraordinary low effective masses. Furthermore, surface steps would act as scattering obstacles and thus reflect standing waves. As a result their wave vector orientation would be perpendicular to the steps, in contrast to our observation of no step correlation (see Fig. 7.12). Thus, standing electron waves can be excluded as origin of the contrast modulation.

7.4 Doping modulation in GaN

The separation of the maxima of the silicon concentration agrees excelently with the distance between neighboring maxima of the contrast modulation. This indicates that the contrast modulation arises from the silicon-doping modulation present in one of the studied GaN epitaxial layers. Therefore, the consequences of a silicon doping modulation will be analyzed in detail. As shown in Fig. 7.16(a) the change of the silicon concentration shifts the energy position of the conduction band edge E_C relative to the Fermi energy E_F. Thereby, with increasing silicon doping concentration the number of states on the GaN surface increases, into which electrons can tunnel at fixed position voltage (shaded regions in Fig. 7.16(b)). This increased current in higher doped layers is compensated by the feedback loop in the constant-current mode by increasing the tip-sample separation. This leads to the measured height modulation in the constant-current STM images in analogy to that of other local band edge changes [212, 213].

The contrast of differently doped areas is also voltage dependent: The number of states, into which electron can tunnel, increases with the voltage V, because all states between E_F and $E_F + eV$ contribute to the tunnel current. Thus, the total current increases with voltage, but the current into the additional states made available for

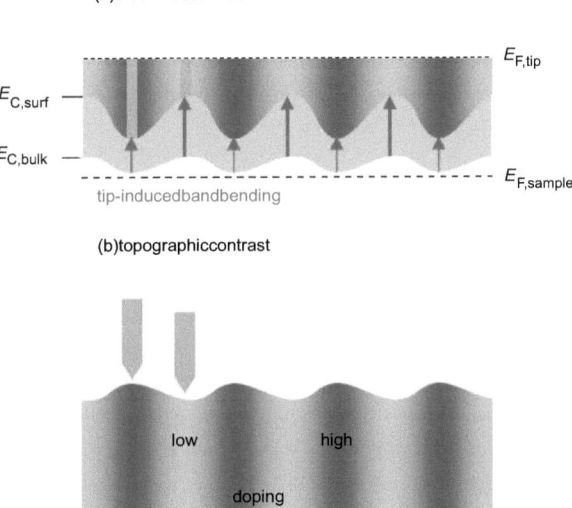

Figure 7.16: *Schematic model of the (a) electronic contrast and (b) the topographic contrast of the observed modulation.*

tunneling because of a local increase of silicon doping (due to modulation doping) remains essentially unchanged. This leads to a decreasing contrast modulation with increasing voltage in agreement with the observation in the inset of Fig. 7.13(a). Similar effects have been observed also on other III-V semiconductor surfaces around charged dopant atoms and defects (screened Coulomb-potential induced band edge shift) [195] and at InAs QDs in GaAs (different QD states) [214].

The strong effect of different tips on the magnitude of the modulation can be explained by taking the additional tip-induced band bending (TIBB) into account, discussed in chapter 3.1. This TIBB leads to a shift of the band edges at the surface relative to those in the bulk (and to E_F), as schematically shown in Fig. 7.16(a). Thus, the number of states available for the tunneling electrons [marked green in Fig. 7.16(a)] decreases with increasing band bending. Thereby the relative fraction of the locally increased doping-induced number of states with respect to the total current and the resulting modulation contrast increase again. Furthermore, the TIBB is doping dependent, being larger at lower doping, which leads to an additional modulation contrast.

The TIBB and thus the modulation are tip dependent. The modulation is largest for a wide, blunt tip, for which the TIBB can be approximated using the one-dimensional model [95, 102]. For an atomically sharp tip, in contrast, the electric field is strongly localized. This leads to a significantly reduced TIBB [97, 105] and therewith to a reduced modulation. Thus, the effect of the TIBB depending on the sharpness of the tip can explain the strong tip sensitivity of the modulation height observed in Fig. 7.13(a).

7.4.4 Theoretical simulation of the topography modulation

The contrast modulation never disappears completely even at very large voltages and for sharp tips. The modulation height is always ≥ 0.4 Å. This non-disappearance at large voltages is in contrast to previous observations of, e.g., charged dopants and defects, where the extended height change induced by the screened Coulomb potential completely disappears at large voltages [195]. However, the silicon doping modulation also induces a mechanical bending of the surface, which is independent of the electronic parameters of the tunneling process [214]. Thereby a constant voltage-independent topographical background modulation is induced [Fig. 7.16(b)], superimposed by electronic effects, discussed above.

A mechanical bending of a surface typically originates from the strain relaxation upon cleavage [214, 215]. In GaN strain can be introduced by silicon doping, because

7.4 Doping modulation in GaN

GaN layers with different silicon doping concentrations are found to exhibit slightly different lattice constants [216]. On basis of these data the lattice constants (a in $[11\bar{2}0]$ and c in [0001] direction) of the lower and higher silicon doped layers present in the sample are calculated [see Fig. 7.13(b)-(c)].

Using silicon concentrations of $n_l = 2.3 \times 10^{18}$ cm^{-3} and $n_h = 3.0 \times 10^{18}$ cm^{-3} as determined from SIMS, lattice constants of $a_l = 3.187084$ Å and $c_l = 5.185937$ Å as well as $a_h = 3.187880$ Å and $c_h = 5.185583$ Å are obtained, respectively [217]. Note, while a increases, c decreases with the silicon doping concentration. M. Winkelnkemper used a slab of 15 periods of alternating 120 nm thick low (n_l) and 150 nm thick high (n_h) silicon-doped GaN layers along the [0001] direction [Fig. 7.13(b)] and the corresponding lattice parameters for the two doping levels as derived above for finite difference method calculations based on continuum mechanics. In this calculation the stiffness constants of Ref. [218] were used. From the calculated strain relaxation the height corrugation of the $(1\bar{1}00)$ surface along the [0001] direction were extracted [Fig. 7.13(c)]. The silicon doping modulation leads to a structural relaxation varying by 0.35 Å independent of the tunneling conditions and in addition to electronic effects. This is in good agreement with the minimum measured modulation height of 0.4 Å. Furthermore, both contrast mechanisms of the silicon doping modulation, strain relaxation and band edge shift, lead to higher tip heights with increasing silicon doping concentration. Thus, both effects add up and have no phase shift, explaining that the modulation does not spatially shift with different voltages or tips.

Chapter 8

Conclusion

In this work different nitrogen containing III-V semiconductor structures are studied using (cross-sectional) scanning tunneling microscopy and spectroscopy. These structures are ternary and quaternary III-V compounds containing only low nitrogen concentrations as well as pure binary GaN, both demonstrating a wide range of interesting structural and electronic properties. Many of these properties were observed within this work for the first time.

First, the incorporation of nitrogen into GaAs is investigated and the structural properties of GaAsN layers grown in GaAs are studied at the atomic scale. In this process several different nitrogen-related features are discussed. The nitrogen arrangement appears to have a random distribution, and nitrogen-related features can be distinguished from other crystal defects such as impurities, vacancies, or interstitials on the base of height profiles from the images. Also, the electronic properties of the GaAsN are studied. The measured local density of states demonstrates that nitrogen impurities drastically change the original GaAs density of states, splitting the conduction band of GaAs. The measured density of states is compared with the band anticrossing model and the density of state calculated using the Green's function approach. In both cases an excellent agreement is demonstrated.

Furthermore, the influence of nitrogen incorporation on the InAs/GaAs(N) quantum dot formation is studied. In the case of nitrogen incorporation into the GaAs matrix of InAs quantum dots two different effects are observed. InAs/GaAs quantum dots capped with GaAsN show equal sizes, as compared with InAs/GaAs quantum dots capped with GaAs. Interestingly, their emission wavelength demonstrates a redshift, which can be attributed to the reduction of the electron confinement by the reduced band gap of the matrix. In contrast, the growth of InAs quantum dots on a GaAsN layer and capped with GaAs has almost no effect on the emission wavelength.

This is due to a strong reduction in their spatial volume, leading to a blue-shift of the emission wavelength compensating the red-shift induced by the change of the confinement due to the nitrogen incorporation.

Moreover, nitrogen incorporation directly into InAs/GaAs quantum dots is investigated. It is observed that nitrogen exposure during InAs growth leads to a rather strong dissolution and the formation of extended almost spherical nitrogen-free InGaAs quantum dots with a low indium content. Instead nitrogen atoms are observed in the surrounding GaAs matrix. Furthermore, indium atoms are found underneath the nominal base plane, where quantum dot growth started, demonstrating strong strain-induced intermixing already during quantum dot formation. It is clearly demonstrated that the InAsN/GaAs quantum dot system shows a trend to separate into InGaAs quantum dots and a GaAsN matrix, the latter one forming due to nitrogen segregation from the quantum dots into the surrounding host material. This indicates that nitrogen incorporation into InAs is suppressed in the case of neighboring GaAs. Due to an incomplete or even missing wetting layer, the quantum dot shape is not flattened during capping, as usually observed for InAs/GaAs quantum dots. In contrast, it develops toward a more spherical shape closer to its thermodynamical equilibrium structure.

In addition to diluted III-V nitrides also the structural and electronic properties of pure n-type epitaxially grown GaN substrates are studied. The identification of the energy positions and types of surface states as well as the origin of the Fermi level pinning on $\text{GaN}(1\bar{1}00)$ cleavage surfaces is presented. Thereby, both the nitrogen and gallium derived dangling bond surface states are found to be outside of the fundamental band gap. Their band edges are both situated at the $\bar{\Gamma}$ point of the surface Brillouin zone. The observed Fermi level pinning at 1 eV below the conduction band edge is attributed to extrinsic states from surface defects such as steps, but not to intrinsic surface states.

Furthermore, the type, the spatial distribution, the projected line directions, and the electronic properties of dislocations in GaN wafers are investigated. The dislocations are found to form localized bunches of entangled nonparallel dislocation lines. Within these bunches uncharged perfect dislocations with $a/3 \langle 11\bar{2}0 \rangle$ Burgers vectors as well as negatively charged Shockley partial dislocations with $a/3 \langle 1\bar{1}00 \rangle$ Burgers vectors and the related intrinsic type-2 stacking fault are found. The observation of charged dislocations suggests that the dissociation of dislocations may be responsible for the insertion of additional gap states in n-type GaN, which may impair the device performance.

Moreover, an epitaxially grown silicon doping modulation structure is imaged. The silicon doping modulation gives rise to a height modulation in the STM images. The origin of the height modulation is traced to two contrast mechanisms, an electronic modulation of the band edge energy yielding a voltage dependent corrugation superposed by a mechanical relaxation at the surface of the doping-induced strain yielding a voltage independent contrast modulation.

It can be concluded that nitrogen containing semiconductor materials exhibit a wide range of interesting structural and electronic properties, allowing to apply them for different types of semiconductor devices. Moreover, the used method, the scanning tunneling microscopy and spectroscopy is an excellent tool to study spatial as well as local electronic properties of such semiconductor materials on the atomic scale. In this work, the ability to study cross-sectional $\mathrm{GaN}(1\bar{1}00)$ surfaces on the atomic scale was demonstrated for the first time. This opens a broad perspective for further studies of nitride-based heterostructures such as InN/GaN quantum dots or GaN/AlN quantum wells.

Bibliography

[1] J. Bardeen and W.H. Brattain, Physical Review **74**, 230 (1948).

[2] B. Gil (Ed.), *Group III Nitride Semiconductor Compounds*, Oxford Science Publications, Physics and Applications, 2008.

[3] P. Schittenhelm, C. Engel, F. Findeis, G. Abstreiter, A.A. Darhuber, G. Bauer, A.O. Kosogov, and P. Werner, Journal of Vacuum Science & Technology B **16**, 1575 (1998).

[4] R.A. Soref, Materials Research Society Bulletin **23**, 20 (1998).

[5] N.H. Bonadeo, J. Erland, D. Gammon, D. Park, D.S. Katzer, and D.G. Steel, Science **282**, 1473 (1998).

[6] S. Adachi, *GaAs and Related Materials: Bulk Semiconducting and Superlattice Properties*, World Scientific, Singapore, 1994.

[7] S. Adachi, *Properties of Group-IV, III-V and II-VI Semiconductors*, Wiley, Chichester, 2005.

[8] H. Eisele, private comunications, 2006.

[9] R. Driad, Z.H. Lu, S. Charbonneau, W.R. McKinnon, S. Laframboise, P.J. Poole, and S.P. McAlister, Applied Physics Letters **73**, 665 (1998).

[10] T. Lundstrom, W. Schoenfeld, H. Lee, and P.M. Petroff, Science **286**, 2312 (1999).

[11] M.C. Bödefeld, R.J. Warburton, K. Karrai, J.P. Kotthaus, G. Medeiros-Ribeiro, and P.M. Petroff, Applied Physics Letters **74**, 1839 (1999).

[12] S. Fafard, Z.R. Wasilewski, C.N. Allen, K. Hinzer, J.P. McCaffrey, and Y. Feng, Applied Physics Letters **75**, 986 (1999).

[13] V.M. Ustinov, E.R. Weber, S. Ruvimov, Z. Liliental-Weber, A.E. Zhukov, A.Yu. Egorov, A.R. Kovsh, A.F. Tsatsul'nikov, and P.S. Kop'ev, Applied Physics Letters **72**, 362 (1998).

[14] N.N. Ledentsov, V.A. Shchukin, M. Grundmann, N. Kirstaedter, J. Böhrer, O. Schmidt, D. Bimberg, V.M. Ustinov, A.Yu. Egorov, A.E. Zhukov, P.S. Kop'ev, S.V. Zaitsev, N.Yu. Gordeev, Zh.I. Alferov, A.I. Borovkov, A.O. Kosogov, S.S. Ruvimov, P. Werner, U. Gösele, and J. Heydenreich, Physical Review B **54**, 8743 (1996).

[15] Y. Lam, J.P. Loehr, and J. Singh, IEEE Journal of Quantum Electronics **28**, 1248 (1992).

[16] F. Capasso, J. Faist, C. Sirtori, and A.Y. Cho, Solid State Communications **102**, 231 (1992).

[17] O.B. Shchekin and D.G. Deppe, Applied Physics Letters **80**, 3277 (2002).

[18] M. Fischer, D. Bisping, B. Marquardt, and A. Forchel, IEEE Photonics Technology Letters **19**, 1030 (2007).

[19] V.V. Mamutin, O.V. Bondarenko, A.Y. Egorov, N.V. Kryzhanovskaya, Y.M. Shernyakov, and V.M. Ustinov, Technical Physics Letters **32**, 229 (2006).

[20] S. Nakamura, M. Senoh, S.-I. Nagahama, N. Iwasa, T. Yamada, T. Matsushita, H. Kiyoku, and Y. Sugimoto, Applied Physics Letters **68**, 3269 (1996).

[21] S. Strite and H. Morkoç, Journal of Vacuum Science & Technology B **10**, 1237 (1992).

[22] P.K. Kandaswamy, F. Guillot, E. Bellet-Amalric, E. Monroy, L. Nevou, M. Tchernycheva, A. Michon, F.H. Julien, E. Baumann, F.R. Giorgetta, D. Hofstetter, T. Remmele, M. Albrecht, S. Birner, and L.S. Dang, Journal of Applied Physics **104**, 093501 (2008).

[23] S. Nakamura, T. Mukai, and M. Senoh, Applied Physics Letters **64**, 1687 (1994).

[24] G. Cywiński, C. Skierbiszewski, A. Fedunieiwcz-Żmuda, M. Siekacz, L. Nevou, L. Doyennette, M. Tchernycheva, F.H. Julien, P. Prystawko, M. Kryśko, S. Grzanka, I. Grzegory, A. Presz, J.Z. Domagała, J. Smalc, M. Albrecht, T. Remmele, and S. Porowski, Journal of Vacuum Science & Technology B **24**, 1505 (2006).

[25] O. Moriwaki, T. Someya, K. Tachibana, S. Ishida, and Y. Arakawa, Applied Physics Letters **76**, 2361 (2000).

[26] R.A. Oliver, G.A.D. Briggs, M.J. Kappers, C.J. Humphreys, and S. Yasin, Applied Physics Letters **83**, 755 (2003).

[27] R. Seguin, S. Rodt, A. Strittmatter, L. Reißmann, T. Bartel, A. Hoffmann, D. Bimberg, E. Hahn, and D. Gerthsen, Applied Physics Letters **84**, 4023 (2004).

[28] H. Schömig, S. Halm, A. Forchel, G. Bacher, J. Off, and F. Scholz, Physical Review Letters **92**, 106802 (2004).

[29] J.H. Rice, J.W. Robinson, J.H. Na, K.H. Lee, R.A. Taylor, D.P. Williams, E.P. O'Reilly, A.D. Andreev, Y. Arakawa, and S. Yasin, Nanotechnology **16**, 1477 (2005).

[30] K. Sebald, H. Lohmeyer, J. Gutowski, T. Yamaguchi, and D. Hommel, Physica Status Solidi (b) **243**, 1661 (2006).

[31] S. Kako, K. Hoshino, S. Iwamoto, S. Ishida, and Y. Arakawa, Applied Physics Letters **85**, 64 (2004).

[32] R. Bardoux, T. Guillet, P. Lefebvre, T. Taliercio, T. Bretagnon, S. Rousset, B. Gil, and F. Semond, Physical Review B **74**, 195319 (2006).

[33] D. Simeonov, A. Dussaigne, R. Butté, and N. Grandjean, Physical Review B **77**, 075306 (2008).

[34] F. Rol, S. Founta, H. Mariette, B. Daudin, L.S. Dang, J. Bleuse, D. Peyrade, J.-M. Gérard, and B. Gayral, Physical Review B **75**, 125306 (2007).

[35] J. Simon, N.T. Pelekanos, C. Adelmann, E. Martinez-Guerrero, R. André, B. Daudin, L.S. Dang, and H. Mariette, Physical Review B **68**, 035312 (2003).

[36] D. Lagarde, A. Balocchi, H. Carrère, P. Renucci, T. Amand, X. Marie, S. Founta, and H. Mariette, Physical Review B **77**, 041304R (2008).

[37] D. Queren, A. Avramescu, G. Brüderl, A. Breidenassel, M. Schillgalies, S. Lutgen, and U. Strauß, Applied Physics Letters **94**, 081119 (2009).

[38] I. Vurgaftman and J.R. Meyer, Journal of Applied Physics **94**, 3675 (2003).

[39] R.W.G Wyckoff (Ed.), *Crystal Structures*, Wiley & Sons, New York, 2nd Ed., 1963.

[40] L. Vegard, Zeitschrift für Physik **5**, 17 (1921).

[41] J.A. Van Vechten and T.K. Bergstresser, Physical Review B **1**, 3351 (1970).

[42] M. Weyers, M. Sato, and H. Ando, Japanese Journal of Applied Physics **31**, L853 (1992).

[43] T. Ahlgren, E. Vainonen-Ahlgren, J. Likonen, W. Li, and M. Pessa, Applied Physics Letters **80**, 2314 (2002).

[44] M. Weyers, M. Sato, and H. Ando, Japanese Journal of Applied Physics **31**, L853 (1992).

[45] K. Onabe, D. Aoki, J. Wu, H. Yaguchi, and Y. Shiraki, Physica Status Solidi A **176**, 231 (1999).

[46] W. Shan, W. Walukiewicz, J.W. Ager III, E.E. Haller, J.F. Geisz, D.J. Friedman, J.M. Olson, and S.R. Kurtz, Physical Review Letters **82**, 1221 (1999).

[47] B. Streetman and S. Banerjee, *Solid State Electronic Devices*, Prentice Hall, New Jersey, 5th Ed., 2000.

[48] I. Vurgaftman, J.R. Meyer, and L.R. Ram-Mohan, Journal of Applied Physics **89**, 5815 (2001).

[49] H. Abu-Farsakh and A. Qteish, Physical Review B **75**, 085201 (2007).

[50] Y. Al-Douri, H. Abid, and H. Aourag, Material Chemistry **65**, 117 (2000).

[51] J.C. Phillips, *Bond and Bands in Semiconductors*, Academis Press, San Diego, 1973.

[52] D. Sasireka, E. Palaniyandi, and K. Iyakutti, Journal of Physics **52**, 81 (1999).

[53] W.A. Harrison, *Electronic Structure and the Properties of Solids, The Physics of the Chemical Bond*, W.H. Freeman and Company, San Francisco, 1980.

[54] J.C. Slater, Journal of Chemical Physics **41**, 3199 (1964).

[55] B. Cordero, V. Gómez, A.E. Platero-Prats, M. Revés, J. Echeverría, E. Cremades, F. Barragán, and S. Alvarez, Dalton Transactions , 2832 (2008).

[56] L. Pauling, Journal of American Chemical Society **54**, 3570 (1932).

[57] L. Bellaiche, S.-H. Wei, and A. Zunger, Physical Review B **54**, 17568 (1996).

[58] J. Wu, W. Shan, and W. Walukiewicz, Semiconductor Science and Technology **17**, 860 (2002).

[59] A. Lindsay, S. Tomić, and E.P. O'Reilly, Solid-State Electronics **47**, 443 (2003).

[60] M. Sopanen, H.P. Xin, and C.W. Tu, Applied Physics Letters **76**, 994 (2000).

[61] S.A. Choulis, T.J.C. Hosea, S. Tomić, M. Kamal-Saadi, A.R. Adams, E.P. O'Reilly, B.A. Weinstein, and P.J. Klar, Physical Review B **66**, 165321 (2002).

[62] C. Skierbiszewski, P. Perlin, P. Wisniewski, W. Knap, T. Suski, W. Walukiewicz, W. Shan, K.M. Yu, J.W. Ager III, E.E. Haller, J.F. Geisz, and J.M. Olson, Applied Physics Letters **76**, 2409 (2000).

[63] K.M. Yu, W. Walukiewicz, W. Shan, J.W. Ager III, J. Wu, E.E. Haller, J.F. Geisz, D.J. Friedman, and J.M. Olson, Physical Review B **61**, R13337 (2000).

[64] Z. Pan, L.H. Li, Y.W. Lin, B.Q. Sun, D.S. Jiang, and W.K. Ge, Applied Physics Letters **78**, 2217 (2001).

[65] A.E. Zhukov, A.R. Kovsh, E.S. Semenova, V.M. Ustinov, L. Wei, J.-S. Wang, and J.Y. Chi, Semiconductors **36**, 899 (2002).

[66] L. Bellaiche, Applied Physics Letters **75**, 2578 (1999).

[67] A. Al-Yacoub and L. Bellaiche, Physical Review B **62**, 10847 (2000).

[68] J.H. Davies, *The Physics of Low-Dimensional Semiconductors, An Introduction*, Cambridge University Press, Cambridge, 1998.

[69] M.A. Herman and H. Sitter, *Molecular Beam Epitaxy - Fundamentals and Current Status*, Springer, Berlin, 1989.

[70] V. Shchukin, N.N. Ledentsov, and D. Bimberg, *Epitaxy of Nanostructures*, Springer, Berlin, 2003.

[71] H. Eisele, *Cross-Sectional Scanning Tunneling Microscopy of InAs/GaAs Quantum Dots*, Wissenschaft & Technik Verlag, Berlin, 2002.

[72] R. Timm, *Formation, Atomic Structure, and Electronic Properties of GaSb Quantum Dots in GaAs*, PhD thesis, TU Berlin, http://opus.kobv.de/tuberlin/volltexte/2008/1731/, 2007.

[73] P.A. Dowben and A. Miller, *Surface Segregation Phenomena*, CRC Press, Boston, 1990.

[74] D. McLean, *Grain Boundaries in Metals*, Oxford University Press, London, 1957.

[75] J.R. Rellick and C.J. McMahon, Metallurgical Transactions **5**, 2439 (1974).

[76] J.M. Moison, C. Guille, F. Houzay, F. Barthe, and M. Van Rompay, Physical Review B **40**, 6149 (1989).

[77] C. Guille, F. Houzay, J.M. Moison, and F. Barthe, Surface Science **189/190**, 1041 (1987).

[78] J.W. Orton and C.T. Foxon, Reports on Progress in Physics **61**, 1 (1998).

[79] D. Hull and D.J. Bacon, *Introduction to Dislocations*, Pergamon Press, Oxford, 3rd Ed., 1984.

[80] S. Amelinck, *Dislocations in Solids*, North-Holland, Amsterdam, 1979.

[81] A.H. Cottrell, *Theory of Crystal Dislocations*, Gordon and Breach, New York, 1964.

[82] I. Grzegory and S. Porowski, Thin Solid Films **367**, 281 (2000).

[83] S. Gu, R. Zhang, Y. Shi, and Y. Zheng, Journal of Physics D: Applied Physics **34**, 1951 (2001).

[84] S. Bohyama, K. Yoshikawa, H. Naoi, H. Miyake, K. Hiramatsu, Y. Iyechika, and T. Maeda, Physica Status Solidi (a) **194**, 528 (2002).

[85] C. Kisielowski, J. Krüger, S. Ruminov, T. Suski, J.W. Ager III, E. Jones, Z. Liliental-Weber, M. Rubin, E.R. Weber, M.D. Bremser, and R.F. Davis, Physical Review B **54**, 17745 (1996).

[86] P. Gibart, Reports on Progress in Physics **67**, 667 (2004).

[87] G. Binnig, H. Rohrer, Ch. Gerber, and E. Weibel, Physical Review Letters **49**, 57 (1982).

[88] G. Binnig and H. Rohrer, Helvetica physica acta **55**, 726 (1982).

[89] R. Shankar, *Principles of Quantum Mechanics*, Springer, New York, 2nd Ed., 1994.

[90] J. Bardeen, Physical Review Letters **6**, 57 (1961).

[91] J. Tersoff and D.R. Hamann, Physical Review Letters **50**, 1998 (1983).

[92] J. Tersoff and D.R. Hamann, Physical Review B **31**, 805 (1985).

[93] J. Chen, *Introduction to Scanning Tunneling Microscopy*, Oxford University Press, New York, 1993.

[94] R.J. Hamers, Annual Review of Physical Chemistry **40**, 531 (1989).

[95] N.D. Jäger, E.R. Weber, K. Urban, and Ph. Ebert, Physical Review B **67**, 165327 (2003).

[96] N.D. Jäger, M. Marso, M. Salmeron, E.R. Weber, K. Urban, and Ph. Ebert, Physikcal Review B **67**, 165307 (2003).

[97] R.M. Feenstra, Physical Review B **50**, 4561 (1994).

[98] R.M. Feenstra, D.A. Collins, D.Z.-Y. Ting, M.W. Wang, and T.C. McGill, Journal of Vacuum Science & Technology B **12**, 2592 (1994).

[99] R.M. Feenstra, J.A. Stroscio, and A.P. Fein, Surface Science **181**, 295 (1987).

[100] S. Gwo, A.R. Smith, K.-J. Chao, C.-K. Shih, K. Sadra, and B.G. Streetman, Journal of Vacuum Science Technology A **12**, 2005 (1994).

[101] H.-A. Lin, R.J. Jaccodine, and M.S. Freund, Applied Physics Letters **74**, 1105 (1999).

[102] R.M. Feenstra and J.A. Stroscio, Journal of Vacuum Science & Technology B **5**, 923 (1987).

[103] R.M. Feenstra and P. Mårtensson, Physical Review Letters **61**, 447 (1988).

[104] P. Mårtensson and R.M. Feenstra, Physical Review B **39**, 7744 (1989).

[105] N.D. Lang, Physical Review Letters **58**, 45 (1987).

[106] D.A. Bonnell (Ed.), *Scanning Tunneling Microscopy and Spectroscopy, Theory, Techniques, and Applications*, VCP Publishers, New York, 1993.

[107] I. Appelbaum, R. Sheth, I. Shalish, K.J. Russell, and V. Narayanamurti, Physical Review B **67**, 155307 (2003).

[108] R.M. Feenstra, Y. Dong, M.P. Semtsiv, and W.T. Masselink, Nanotechnology **18**, 044015 (2007).

[109] R. Dombrowski, Ch. Steinebach, Ch. Wittneven, M. Morgenstern, and R. Wiesendanger, Physical Review B **59**, 8043 (1999).

[110] H.-A. Lin, R.J. Jaccodine, and M.S. Freund, Applied Physics Letters **73**, 2462 (1998).

[111] S.M. Sze, *Physics of Semiconductor Devices*, John Wiley & Sons, New York, 2nd Ed., 1981.

[112] J.L. Chelikowsky and M.L. Cohen, Physical Review B **20**, 4150 (1979).

[113] S. Borisova, *Investigation of Broad Band Gap Semiconductors using Scanning Tunneling Microscopy*, Master thesis, Lomonosov Moscow State University, 2008.

[114] J.E. Northrup and J. Neugebauer, Physical Review B **53**, R10477 (1996).

[115] C.G. Van de Walle and D. Segev, Journal of Applied Physics **101**, 081704 (2007).

[116] D.M. Ceperley and B.J. Alder, Physical Review Letters **45**, 566 (1980).

[117] B. Siemens, C. Domke, Ph. Ebert, and K. Urban, Physical Review B **56**, 12321 (1997).

[118] P. Schröder, P. Krüger, and J. Pollmann, Physical Review B **49**, 17092 (1994).

[119] A. Lenz, *Atomic Structure of Capped In(Ga)As and GaAs Quantum Dots for Optoelectronic Devices*, PhD thesis, TU Berlin, http://opus.kobv.de/tuberlin/volltexte/2008/1772/, 2008.

[120] J.P. Ibe, P.P. Bey, Jr., S.L. Brandow, R.A. Brizzolara, N.A. Burnham, D.P. DiLella, K.P. Lee, C.R.K. Marrian, and R.J. Colton, Journal of Vacuum Science & Technology A **8**, 3570 (1990).

[121] H.P. Xin and C.W. Tu, Applied Physics Letters **72**, 2442 (1998).

[122] I.A. Buyanova, W.M. Chen, G. Pozina, J.P. Bergman, B. Monemar, H.P. Xin, and C.W. Tu, Applied Physics Letters **75**, 501 (1999).

[123] W. Li, M. Pessa, J. Toivonen, and H. Lipsanen, Physical Review B **64**, 113308 (2001).

[124] O. Schumann, S. Birner, M. Baudach, L. Geelhaar, H. Eisele, L. Ivanova, R. Timm, A. Lenz, S.K. Becker, M. Povolotskyi, M. Dähne, G. Abstreiter, and H. Riechert, Physical Review B **71**, 245316 (2005).

[125] O. Schumann, L. Geelhaar, H. Riechert, H. Cerva, and G. Abstreiter, Journal of Applied Physics **96**, 2832 (2004).

[126] H.A. McKay, R.M. Feenstra, T. Schmidtling, U.W. Pohl, and J.F. Geisz, Journal of Vacuum Science & Technology B **19**, 1644 (2001).

[127] H.A. McKay, R.M. Feenstra, T. Schmidtling, and U.W. Pohl, Applied Physics Letters **78**, 82 (2001).

[128] J.M. Ulloa, P.M. Koenraad, and M. Hopkinson, Applied Physics Letters **93**, 083103 (2008).

[129] S. Rubini, G. Bais, A. Cristofoli, M. Piccin, R. Duca, C. Nacci, S. Modesti, E. Carlino, F. Martelli, A. Franciosi, G. Bisognin, D. De Salvador, P. Schiavuta, M. Berti, and A.V. Drigo, Applied Physics Letters **88**, 141923 (2006).

[130] H. Abu-Farsakh and J. Neugebauer, Physical Review B **79**, 155311 (2009).

[131] G. Schwarz, *Untersuchungen zu Defekten auf und nahe der (110) Oberfläche von GaAs und weiteren III-V-Halbleitern*, PhD thesis, TU Berlin, http://edocs.tu-berlin.de/diss/archiv_2002.html, 2002.

[132] Ph. Ebert, P. Quadbeck, K. Urban, B. Henninger, K. Horn, G. Schwarz, J. Neugebauer, and M. Scheffler, Applied Physics Letters **79**, 2877 (2001).

[133] M. Prietsch, A. Samsavar, and R. Ludeke, Physical Review B **43**, 11850 (1991).

[134] R.S. Goldman, B.G. Briner, R.M. Feenstra, M.L. O'Steen, and R.J. Hauenstein, Applied Physics Letters **69**, 3698 (1996).

[135] J. Wu, W. Walukiewicz, and E.E. Haller, Physical Review B **65**, 233210 (2002).

[136] M.P. Vaughan and B.K. Ridley, Physical Review B **75**, 195205 (2007).

[137] F. Heinrichsdorff, M.-H. Mao, N. Kirstaedter, A. Krost, and D. Bimberg, Applied Physics Letters **71**, 22 (1997).

[138] M. Grundmann, Physica E **5**, 167 (2000).

[139] A. Lenz, H. Eisele, R. Timm, L. Ivanova, H.-Y. Liu, M. Hopkinson, U.W. Pohl, and M. Dähne, Physica E **40**, 1988 (2008).

[140] A. Lenz, H. Eisele, R. Timm, L. Ivanova, R.L. Sellin, H.-Y. Liu, abd U.W. Pohl M. Hopkinson, D. Bimberg, and M. Dähnee, Physica Status Solidi (b) **246**, 717 (2009).

[141] J.F. Chen, C.H. Yang, Y.H. Wu, L. Chang, and J.Y. Chi, Journal of Applied Physics **104**, 103717 (2008).

[142] Z.Z. Sun, S.F. Yoon, K.C. Yew, and B.X. Bo, Journal of Crystal Growth **263**, 99 (2004).

[143] T. Hakkarainen, J. Toivonen, M. Sopanen, and H. Lipsanen, Journal of Crystal Growth **248**, 339 (2003).

[144] F. Guffarth, R. Heitz, A. Schliwa, O. Stier, N.N. Ledentsov, A.R. Kovsh, V.M. Ustinov, and D. Bimberg, Physical Review B **64**, 085305 (2001).

[145] F. Guffarth, R. Heitz, A. Schliwa, O. Stier, A.R. Kovsh, V. Ustinov, N.N. Ledentsov, and D. Bimberg, Physica Status Solidi (b) **224**, 61 (2001).

[146] V. Holý, G. Springholz, M. Pinczolits, and G. Bauer, Physical Review Letters **83**, 356 (1999).

[147] O. Flebbe, H. Eisele, T. Kalka, F. Heinrichsdorff, A. Krost, D. Bimberg, and M. Dähne-Prietsch, Journal of Vacuum Science & Technology B **17**, 1639 (1999).

[148] T. Ahlgren, E. Vainonen-Ahlgren, J. Likonen, W. Li, and M. Pessa, Applied Physics Letters **80**, 2314 (2002).

[149] A.M. Mintairov, P.A. Blagnov, V.G. Melehin, N.N. Faleev, J.L. Merz, Y. Qiu, S.A. Nikishin, and H. Temkin, Physical Review B **56**, 15836 (1997).

[150] R.J. Kaplar, S.A. Ringel, S.R. Krutz, J.F. Klem, and A.A. Allerman, Applied Physics Letters **80**, 4777 (2002).

[151] P. Krispin, S.G. Spruytte, J.S. Harris, and K.H. Ploog, Journal of Applied Physics **88**, 4153 (2000).

[152] H. Eisele, R. Timm, A. Lenz, Ch. Hennig, M. Ternes, S.K. Becker, and M. Dähne, Physica Status Solidi (c) **0**, 1129 (2003).

[153] A. Lenz, R. Timm, H. Eisele, Ch. Hennig, S.K. Becker, R.L. Sellin, U.W. Pohl, D. Bimberg, and M. Dähne, Applied Physics Letters **81**, 5150 (2002).

[154] A. Lenz, H. Eisele, R. Timm, S.K. Becker, R.L. Sellin, U.W. Pohl, D. Bimberg, and M. Dähne, Applied Physics Letters **85**, 3848 (2004).

[155] M. Albrecht, H. Abu-Farsakh, T. Remmele, L. Geelhaar, H. Riechert, and J. Neugebauer, Physical Review Letters **99**, 206103 (2007).

[156] W.M. McGee, R.S. Williams, M.J. Ashwin, and T.S. Jones, Surface Science **600**, L194 (2006).

[157] Th. Hammerschmidt, *Growth Simulations of InAs/GaAs Quantum-Dots*, PhD thesis, TU Berlin, http://opus.kobv.de/tuberlin/volltexte/2006/1358/, 2006.

[158] R.S. Williams, W.M. McGee, M.J. Ashwin, T.S. Jones, E. Clarke, P. Stavrinou, J. Zhang, S. Tomić, and C.P.A. Mulcahy, Applied Physics Letters **90**, 032109 (2007).

[159] H. Eisele, A. Lenz, Ch. Hennig, R. Timm, M. Ternes, and M. Dähne, Journal of Crystal Growth **248**, 322 (2003).

[160] L. Ivanova, H. Eisele, A. Lenz, R. Timm, M. Dähne, O. Schumann, L. Geelhaar, and H. Riechert, Applied Physics Letters **92**, 203101 (2008).

[161] I. Daruka, J. Tersoff, and A.-L. Barabási, Physical Review Letters **82**, 2753 (1999).

[162] J.F. Nye (Ed.), *Physical Properties of Crystals*, Oxford University Press, Oxford, 1985.

[163] D. Feezell, S. Nakamura, S. DenBaars, and J. Speck, *Short-Wave Diode Lasers: Nonpolar Gallium Nitride Laser Diodes are the next new blue*, Laser Focus World, 2008.

[164] S. Chichibu, T. Azuhata, T. Sota, and S. Nakamura, Applied Physics Letters **69**, 4188 (1996).

[165] P. Waltereit, O. Brandt, A. Trampert, H.T. Grahn, J. Menninger, M. Ramsteiner, M. Reiche, and K.H. Ploog, Nature **406**, 865 (2000).

[166] S.H. Park, Japanese Journal of Applied Physics **42**, L170 (2003).

[167] D.G. Deppe, Applied Physics Letters **56**, 370 (1990).

[168] M.D. Pashley and K.W. Haberern, Physical Review Letters **67**, 2697 (1991).

[169] A.R. Smith, R.M. Feenstra, D.W. Greve, M.S. Shin, M. Skowronski, J. Neugebauer, and J.E. Northrup, Journal of Vacuum Science & Technology B **16**, 2242 (1998).

[170] C. Noguez, Physical Review B **62**, 2681 (2000).

[171] A. Filippetti, V. Fiorentini, G. Cappellini, and A. Bosin, Physical Review B **59**, 8026 (1999).

[172] J. Wichert, R. Weber, L. Kipp, M. Skibowski, T. Strasser, F. Starrost, C. Solterbeck, W. Schattke, T. Suski, I. Grzegory, and S. Porowski, Physical Status Solidi (b) **215**, 751 (1999).

[173] R.M. Feenstra, J.A. Stroscio, J. Tersoff, and A.P. Fein, Physical Review Letters **58**, 1192 (1987).

[174] L. Ivanova, S. Borisova, H. Eisele, M. Dähne, A. Laubsch, and Ph. Ebert, Applied Physics Letters **93**, 192110 (2008).

[175] J.A. Stroscio, R.M. Feenstra, and A.P. Fein, Physical Review Letters **57**, 2579 (1986).

[176] C.J. Chen and R.J. Hamers, Journal of Vacuum Science & Technology B **9**, 503 (1991).

[177] Ph. Ebert, L. Ivanova, H. Eisele, and M. Dähne, Physical Review B accepted (2009).

[178] S.C. Jain, M. Willander, J. Narayan, and R. Van Overstraeten, Journal of Applied Physics **87**, 965 (2000).

[179] J. Cai and F.A. Ponce, Physica Status Solidi (a) **192**, 407 (2002).

[180] G. Koley and M.G. Spencer, Applied Physics Letters **78**, 2873 (2001).

[181] P.J. Hansen, Y.E. Strausser, A.N. Erickson, E.J. Tarsa, P. Kozodoy, E.G. Brazel, J.P. Ibbetson, U. Mishra, V. Narayanamurti, S.P. DenBaars, and J.S. Speck, Applied Physics Letters **72**, 2247 (1998).

[182] J.W.P. Hsu, H.M. Ng, A.M. Sergent, and S.N.G. Chu, Applied Physics Letters **81**, 3579 (2002).

[183] M. Albrecht, L.J. Weyher, B. Lucznik, I. Grzegory, and S. Porowski, Applied Physics Letters **92**, 231909 (2008).

[184] A. Krtschil, A. Dadgar, and A. Krost, Applied Physics Letters **82**, 2263 (2003).

[185] D. Cherns and C.G. Jiao, Physical Review Letters **87**, 205504 (2001).

[186] J. Elsner, R. Jones, P.K. Sitch, V.D. Perzag, M. Elstner, Th. Frauenheim, M.I. Heggie, S. Öberg, and P.R. Briddon, Physical Review Letters **79**, 3672 (1997).

[187] Y. Xin, S.J. Pennycook, N.D. Browning, P.D. Nellist, S. Sivananthan, F. Omnès, B. Beaumont, J.P. Faurie, and P. Gibart, Applied Physics Letters **72**, 2680 (1998).

[188] A.F. Wright and U. Grossner, Applied Physics Letters **73**, 2751 (1998).

[189] A.T. Blumenau, J. Elsner, R. Jones, M.I. Heggie, S. Öberg, T. Frauenheim, and P.R. Briddon, Journal of Physics: Condensed Matter **12**, 10223 (2000).

[190] I. Arslan, A. Bleloch, E.A. Stach, and N.D. Browning, Physical Review Letters **94**, 025504 (2005).

[191] L. Lymperakis, J. Neugebauer, M. Albrecht, T. Remmele, and H.P. Strunk, Physical Review Letters **93**, 196401 (2004).

[192] A. Sakai, H. Sunakawa, and A. Usui, Applied Physics Letters **71**, 2259 (1997).

[193] D.N. Zakharov, Z. Liniental-Weber, B. Wagner, Z.J. Reitmeier, E.A. Preble, and R.F. Davis, Physical Review B **71**, 235334 (2005).

[194] Ph. Ebert, L. Ivanova, S. Borisova, H. Eisele, A. Laubsch, and M. Dähne, Applied Physics Letters **94**, 062104 (2009).

[195] Ph. Ebert, Surface Science Reports **33**, 121 (1999).

[196] C. Domke, M. Heinrich, Ph. Ebert, and K. Urban, Journal of Vacuum Science & Technology B **16**, 2825 (1998).

[197] M. Heinrich, C. Domke, Ph. Ebert, and K. Urban, Physical Review B **53**, 10894 (1996).

[198] G. Savini, A.T. Blumenau, M.I. Heggie, and S. Öberg, Physica Status Solidi (c) **4**, 2945 (2007).

[199] Ph. Ebert, C. Domke, and K. Urban, Applied Physics Letters **78**, 480 (2001).

[200] H. Amano, M. Kitoh, K. Hiramatsu, and I. Akasaki, Journal of Electrochemical Society **137**, 1639 (1990).

[201] A. Krost und A. Dadgar, Materials Science and Engineering B **93**, 77 (2002).

[202] M. Winkelnkemper, A. Schliwa, and D. Bimberg, Physical Review B **74**, 155322 (2006).

[203] O. Albrektsen, D.J. Arent, H.P. Meier, and H.W.M. Salemink, Applied Physics Letters **57**, 31 (1990).

[204] R.M. Feenstra, Semiconductor Science and Technology **9**, 2157 (1994).

[205] R.M. Feenstra, J.M. Woodall, and G.D. Pettit, Physical Review Letters **71**, 1176 (1993).

[206] J.F. Zheng, X. Liu, N. Newman, E.R. Weber, D.F. Ogletree, and M. Salmeron, Physical Review Letters **72**, 1490 (1994).

[207] K.-J. Chao, C.-K. Shih, D.W. Gotthold, and B.G. Streetman, Physical Review Letters **79**, 4822 (1997).

[208] N.D. Jäger, K. Urban, E.R. Weber, and Ph. Ebert, Applied Physics Letters **82**, 2700 (2003).

[209] T.C.G. Reusch, M. Wenderoth, L. Winking, N. Quaas, and R.G. Ulbrich, Physical Review Letters **93**, 206801 (2004).

[210] M.F. Crommie, C.P. Lutz, and D.M. Eigler, Nature **363**, 524 (1993).

[211] O. Flebbe, H. Eisele, R. Timm, and M. Dähne, AIP Conference Proceedings **696**, 699 (2003).

[212] J.A. Stroscio, R.M. Feenstra, and A.P. Fein, Physical Review Letters **58**, 1668 (1987).

[213] R.J. Hamers, Journal of Vacuum Science and Technology B **6**, 1462 (1988).

[214] H. Eisele, O. Flebbe, T. Kalka, and M. Dähne-Prietsch, Surface and Interface Analysis **27**, 537 (1999).

[215] H. Chen, R.M. Feenstra, R.S. Goldman, C. Silfvenius, and G. Landgren, Applied Physics Letters **72**, 1727 (1998).

[216] L.T. Romano, C.G. van de Walle, J.W. Ager III, W. Götz, and R.S. Kern, Journal of Applied Physics **87**, 7745 (2000).

[217] H. Eisele, L. Ivanova, S. Borisova, M. Dähne, M. Winkelnkemper, and Ph. Ebert, Applied Physics Letters **94**, 162110 (2009).

[218] A. Polian, M. Grimsditch, and I. Grzegory, Journal of Applied Physics **79**, 3343 (1996).

Acknowledgments

I would like to acknowledge Prof. Dr. Mario Dähne for the opportunity to study, research, teach, and write my thesis in his group. Thank you very much for your patience and enthusiasm for physics. In particular, I thank you for allowing me great latitude to emerge myself. I am grateful to Priv.-Doz. Dr. Philipp Ebert for refereeing this thesis: he added to a considerable part of the achievements reported in this work. Thank you very much, you are a great partner for a cooperation. Prof. Dr. Erwin Sedlmayr is cordially acknowledged for chairing the final defense. In addition, I very much acknowledge the motivation and the kindliness of Prof. Dr. Hans-Eckhart Gumlich.

My special thanks go to Dr. Holger Eisele, without his contributions my work would have evolved differently. Thank you for all your support: your great motivation, creative ideas, encouragement, fruitful discussions, and the time we have spent together. Thank you for letting me know that you trust in me.

My deepest thanks go to all the former and present members of our group for creating an affectionate atmosphere. I want to thank Dr. Andea Lenz, Dr. Rainer Timm, Matthias Müller, Dr. Mandy Baudach, and Svetlana Borisova for their assistance in taking XSTM measurements. Vivien Voßebürger and Dominik Martin, as well Nadine Oswald I would like to thank for working on the same topic and many discussions. I acknowledge Dr. Ernst Lenz for his analysis software. My special thanks go to Jan Grabowski for many years in the same office, in particular for his helpfulness, ability to speak about every topic, have fun, and for being a friend. I am deeply grateful to Dr. Andrea Lenz, her readiness to help others is amazing. Thank you for your support and your love, dear friend. Dr. Martina Wanke I thank for her stability, optimism, understanding, and very nice time in the office, I miss you. Dr. Kai Hodeck is acknowledged for discussions about health, dance and theater. I would like to thank Gerd Pruskil, as well as Florian Genz, Jonas Becker, as well as Christopher Prohl, Martin Franz, Britta Höpfner, and Matthias Vetterlein for creating a great atmosphere.

My special thanks belong to Prof. Dr. Chih-Kang Ken Shih for his lust for life and the interesting time which I spent in his group in Austin. Furthermore, I would like to thank Dr. Sebastien Founta, Jisun Kim, Linda and Dr. William Hallidy, as well as Jack Clifford for their support during my time in Austin.

I want to thank Dr. Lutz Geelhaar, Dr. Oliver Schumann, and Prof. Dr. Henning Riechert for an effective collaboration, and especial Lutz Geelhaar for his thoroughness. Prof. Dr. Eoin O'Reilly and Dr. Andreas Amann are acknowledged for a short but very fruitful collaboration and insightful discussions about GaAsN. Dr. Momme Winkelnkemper I would like to thank for performing the strain-relaxation calculation and Dr. Sven Rodt for making SEM images.

I want to thank all my friends for being there for me. I want to express my most sincere thanks to my family for rendering every assistance, love, and support.

This work was supported by the Deutsche Forschungsgemeinschaft in the collaborate research center Sfb 296, 787, and project Ei788/1-1.

I want morebooks!

Buy your books fast and straightforward online - at one of the world's fastest growing online book stores! Environmentally sound due to Print-on-Demand technologies.

Buy your books online at
www.get-morebooks.com

Kaufen Sie Ihre Bücher schnell und unkompliziert online – auf einer der am schnellsten wachsenden Buchhandelsplattformen weltweit!
Dank Print-On-Demand umwelt- und ressourcenschonend produziert.

Bücher schneller online kaufen
www.morebooks.de

OmniScriptum Marketing DEU GmbH
Heinrich-Böcking-Str. 6-8
D - 66121 Saarbrücken
Telefax: +49 681 93 81 567-9

info@omniscriptum.com
www.omniscriptum.com

Printed by Books on Demand GmbH, Norderstedt / Germany